Man Remakes

The Earth

or

Two for One

by

S. MERRY

ISBN: 978-1-84728-596-6

Publisher: Lulu.com

Rights Owner: Sidney Merry

Copyright © 2006 by Sidney Merry. All rights reserved.

The author and publisher has done his best to ensure the accuracy of all information delivered but would be pleased to accept any advices to amend any mistakes.

CHAPTERS

INTRODUCTION

THE GREENHOUSE EFFECT

THE OZONE LAYER

FEEDING THE HUNGRY

THE DEVELOPMENT OF RESOURCES

WATER

WASTE DISPOSAL

ACID RAIN

CHEMICALS

ENERGY

NUCLEAR ENERGY AND RADIATION

CONCLUSION

Man Remakes The Earth

Introduction

We are creatures who know life in the form of complex, not fully unravelled, chains of reactions based on carbon chemistry and life itself is set on a fulcrum basis, death occurring when the equilibrium is too far unbalanced. These reactions operate best within narrow margins, and even within these acceptable margins there are fringe conditions and others offering maximum condition for viability. We range in temperature conditions from the Greenland to Sudan and a rise of 2 degrees can improve life and inflict suffering elsewhere.

We also need an intake of food, some of which may best be grown in conditions not available locally. Here we have had to compromise. Securing viability in one direction can bring in its train vitiating, even dangerous side-effects not envisaged in any original picture. Especially does this possibility rear its head when the side effects only manifest themselves at certain levels of concentration, and polarisation develops.

Our adaptability which includes our social adaptations mark us out as the most artificial of beings, and the most successful. We wander everywhere and we do not need to carry a protective shell with us. We can erect one almost anywhere. But we need access to food, clothing and shelter as well as sources of energy.

We exist on a planet where the earth on which we are based moves under the pressures and releases of molten rocks below, and we live with the consequences. Such movements of titanic impact for us, and the movements of the oceans are still little understood and their inclusion in the scheme of things is subject to misrepresentation, misunderstanding, omission despite the fact that they seem to be far more important than the changes in say atmospheric carbon dioxide levels.

We are not about to neglect the contribution of man's contribution but we do wish to indicate our perspective.

Coupled are the omnipresent fears into which our culture blends such facts with doomsday forecasts abounding in all religions but very important in Christianity. Even in the short term we are fed the belief that pleasure and certainly luxury, needs to be paid for by the visitation of a punishing major or minor misfortune, social or personal.

Such programs have been declared with vehemence, by pressure groups seeking wide support on this age-old appeal. Today they are armed with scientific data and certainly were a quarter of their forecasts likely, there should be a demand for large-scale and immediate concern and attention.

We can hardly claim that the effect of pumping greenhouse gases into the atmosphere has no effect, but the sun is giving us an increase in temperature,, the position of Earth with respect to the sun in its orbit will have its effect as will volcanic eruptions. It is however legitimate and most important to ask whether man-made efforts are important in comparison, even if only as the straw that broke the camels back. It is also legitimate to ask whether it is a BAD for man, and just as important how far one could use investigation for the possibility of controlling any threat. Alternatively we need to know how far we can ameliorate its worse consequences.

Four billion years ago, the world had water, carbon dioxide, ammonia and methane. The methane, the carbon dioxide and water vapour trapped the sun's rays and "our" prospect of life commenced when a particular strain of bacteria developed an ability to convert carbon dioxide to oxygen. When the temperature rose, the conversion grew and plant forms developed.

There are to be found inland, seaports of the past. There is to be found agricultural land now covered by the sea and all about there is the movement of land up and down as evidenced by the chalk cliffs about us. Our land floats on a sea of molten rock. In the U.K the south is tilting down relative to Scotland increasing the effect of increased levels of ocean water in that region.

In the pre-Cambrian period 800- 550 million years ago glaciers existed at the tropics. Seventy million years ago Alberta had prolific vegetation, pre-historic reptiles and could even support the tyrannosaurus. Indeed forest life and dinosaur life existed in Antarctica 100 million years ago according to the "Walking with Dinosaurs" series on BBC1 in 1999.

The world today is between Ice Ages and few have turned their mind to the possible disadvantages of a return to an inevitable Ice Age. There is nothing attractive in such an occurrence. The last Ice Age saw snow that never melted in Europe or North America. The causes of these changes are largely unknown and the earth has suffered warmer and colder climates than exist today. We have ascribed these to the gods, shifting sea levels, volcanic activity, comets, changes in the orbital movement of the earth and combinations. More recently another suggestion offered for the end of an ice era has been the threat of a catastrophic release of the abundant methane stored within the earth or tied in the oceans causing a "green" effect. The deposits in question are posited as totalling more than the total fuel reserves so far known from all other sources.

Man Remakes The Earth

The world has suffered an ice age every 100,000 years. There have been changes in the elliptical orbit and a corresponding change in the amount of sunlight falling on earth. A preliminary finding by Dr. Appleton of the Rutherford Laboratories when examining findings arising from the 1999 total eclipse was that the sun was getting hotter than expected. If the sun gets hotter so too will the earth.

Since known time there have been 5 mass extinctions, 248 catastrophic volcanic eruptions as well as numerous tectonic movements causing earthquakes and tidal waves and there is also to mention 65 visits by large asteroids. Indeed Britain has floated greatly from its original position from nearer the equator." "The Independent" of 8/9/99 offers an alternative to the theory that the Sahara and Arabian deserts arose from overgrazing or farming. It suggests that some 10,000 years ago because of inner earth movements the tilt of the earth changed from 24.14 degrees of vertical to 23.45 off vertical, sufficient to change the time when this planet was nearest the sun from July to January. In the plethora of catastrophe theories the most important factor to hold onto is that to date man's worst efforts are at best the straw that could break the camel's back. Even that needs to be taken with a large pinch of salt.

Two hundred million years ago, in the late Triassic period, the carbon dioxide levels in the air increased fourfold without contribution by man. The carbon dioxide level in the air during the last Ice Age was some 2/3rd of that in 1700 and 25,000 years ago when carbon dioxide was 25% lower, with methane 50% lower, the overall world temperature was some 5 degrees centigrade lower. Since 1700 a further 280 billion tons of carbon dioxide has been added to the total increasing the level by 50%. However 70 million years ago there was more in the air than now, and without man's contributions. Without its atmosphere the average temperature on earth would diminish to minus 23 degrees centigrade against present overall levels of 15 degrees. This important difference of 38 degrees arises from atmospheric methane, carbon dioxide and water vapour and there is a contribution from the heat generated within the earth, arising largely from nuclear activity. It is enough to maintain a temperature well above zero.

As plants began to develop their photosynthesising abilities, they spread with the increasing atmospheric carbon dioxide, the latter began gradually to reduce in level. The level further reduced and an Ice Age entered lasting some 100,000 years. It is still a moot point how far the gradual onset of cooler conditions was due to extra-terrestrial conditions, or the absorption of carbon dioxide by an over-expanding vegetation belt. Both could have played a part.

In the Ice Age the gas was present at a level of 180 ppm. By the nineteenth century it had increased by 50% and it is difficult to see here any such contribution by man. The decimation of European forest would hardly explain it. Seas in areas receded, lands rose and fell.

There are estimates that were the level of the gas to rise from its present level at 350 ppm to 560 ppm then temperatures could rise between 1.5 and 4.5 C. From this there arises the forecast that the sea will rise and New York, London, Peking and Holland could face problems. If the water be lost from the Arctic regions, then the land previously weighed down by the weight of that ice will rise. 10,000 years ago the temperature on the Greenland ice cap rose 5 C over 20 years and there was a warm period to 7,000 B.C.. There were lakes in the Sahara, hot summers in the north and cooler summers at the equator.

Again the matter would appear too simplified, since we all know of ports left behind by receding waters, sometimes 3 miles from their present reaches. This informs us only that the movement of land masses, like the role of the oceans and their currents is not that well understood. Yet these vast tectonic movements of masses could materially diminish the importance of factors in use for present day projections. In any case there is no clear connection between temperature changes and the creation of storms. Differential pressures are a direct source and here the oceans must play a major role though again one not adequately understood. (B.B.C. radio 4 18/9/01). Especially in these vast movements there is likely to arise a considerable time lag between cause and effect and 50 years, 100 years are not beyond such sequences.

Farman, who first made us aware of the ozone problem has recorded sea levels for the last 30 years on the Arctic shores. When reinforced with other data he sees no change in the sea level over the last 80 years. He adds that the past decade has witnessed the warmest of weathers and the coldest of winters. "The Independent" of 7/4/96 tells of the Northern ice cap melting through global warming or the disruption in flow of sea currents including the Gulf Stream and these may be interconnected. "The Independent" of 19/1/02 warns again of an increase in the Antarctic by 27 billion tons of ice not a forecast of 21 billion down. On 25/1/96 it reported that 8,000 square kms of ice shelf in the Antarctic were disintegrating and the British Antarctic Survey team were indicating a rise of 2.5 C over the past 50 years but were not sure whether this was a natural oscillation. The Met. Office forecasts a 6-8 C rise in the next century but A Cambridge University view sees the ice melting without rise in sea levels.

It is hard to explain in terms of pollutants why 1106 A.D. gave the second warmest summer to 1998. There was a mini-ice ages between 1350 and 1900 when the Thames froze regularly and the difference in overall temperature was only one degree. Thus small differences can make major contributions and should not be excluded from consideration. Towards the end of the 16th century there was a mini ice-age and in 1596 perpetual rain and near famine. In 1600 A.D. and now associated with a Peruvian volcanic eruption there were frosts in G.B. every morning of June and July. We just have to develop adequate systems of meaningful sensitivity and range! After all this could be the explanation for the sudden change of the herring track from the Baltic to the Atlantic, which brought the English fisherman to the fore with rich harvests. Such changes may also have contributed to crop failures and to the weakening or decline of the dependent populations.

For the past 2 million years Ice Ages have come and gone. The sun is the beginning of it all, although a most important contribution has come from the heat, the sulphur, the carbon derived from below the earth and sea.

The sun gives heat to the equator and less heat to the poles and there results heat transfer affecting the atmosphere. The spinning earth affects the condition of transfer differently in different places. The currents interact with the ocean streams and these have their own time dependent heat exchange systems. The sun's heat changes every 11 years by a little, sun spots make their contribution varied in time through their energy output as received on earth, but the juxtaposition of earth and sun changes may have an even greater effect. The tilt changes the exposure to the sun, there are changes in the axes of the earth and there are elliptical orbit changes. Certainly it is accepted that it is these that have largely determined the ice and warmer ages of the past. Today "Scientist American avers that the movement of the waters around the earth has affected the earth's spin ,if only slightly. Cycles have been derived of 23,000 years, 41,000 and 100,000 each overlapping. The last warm age was 100,000 years ago and there has been much more earth-time spent in ice-ages than otherwise, and even a carbon dioxide level of 180 ppm does not sufficiently explain what happened.

In any ecological evaluation, cause and effect can be separated by 50 years or much more, and little account is being taken of such possible time lags.

There is equilibrium movement between the carbon dioxide in water and air. When the temperature rises the gas is expelled. Rising warmth lends itself to an extension in ocean life and thus an absorption of the additional gas released to the atmosphere. During the ice age, less methane was available. Earth, sea and air are the necessary sources of our viability. Powered by the sun, waste enters these arenas and this waste is the base source for conversion to life. The sustenance of one is the waste of another. Sometimes the relation can be seen as with the dung beetle. Alternatively it can involve the complex cycles of nitrogen, oxygen and carbon.

Even in existing conditions, including factors from moon and sun there is change. That change is consequent on the interaction of such equilibria. It may well be the case that the time spans involved in the juxtaposition of all these variables denotes change over enormous periods so that what we consider as the desirable status quo, may change to a new equilibrium, inimical to ourselves in absolute terms or in favouring the development of competition. Only knowledge used competently could ever help in such circumstances.

The Greenhouse Effect

Our information tells us that in 1765 the carbon dioxide content of the atmosphere was 280 ppm and that it is now 350, that methane has doubled to 1.7 ppm, that nitrogen oxide gases has risen about 10% to 300 ppb and that as a result there has been a rise in average world temperature of about 0.5%. There is also the relevant point that thermal inertia will slow the onset of temperature rises but not necessarily reduce its level. The additional cry is that the immediacy of the vast projected changes may exacerbate the position in that life might not adapt fast enough. Yet other sources already tell us that the last Ice Age ended some 15,000 years ago and within 300 years of its end G.B. was a temperate zone with corresponding vegetation. It is also the case that in the era of the Dinosaur the land was lush and it serviced them too.

It goes without saying that we cannot control the movement of the Earth in relation to its non-circular orbit around the sun. We cannot control the fact that the sun is increasing in temperature, nor can we control the earthquakes, the volcanoes from which there have been more than 3,200 eruptions between 1700 and 2005 or the El Ninos. Our purpose in this essay is to persuade that such phenomena are far more important than man's efforts. It may well be that the additional contribution of man could become the straw to break the camels back, but it is desirable to bear in mind that the level of carbon dioxide in the era of the Dinosaur was higher than today and they did not burn carbon fuels. It is also important to bear in mind that water moisture however produced has a greenhouse effect. Ruddiman suggests that methane began to increase with the production of rice, some 5,000 years ago. Yet how can we even consider curtailing agriculture or the expansion of domestic animals when man's needs have manifestly not yet been met?

Inevitably we are on our way to an ice age yet the Luddite in us wants no change. Nevertheless should we not act more positively in order to lessen its possible impact on our future progeny. Should we not use any possibility of ameliorating the coming ice age?

Certainly additional greenhouse gases must mean warming but it is our aim to show that present "facts" as presented are little short of speculation and wild speculation at that. The Economist of 2/4/05 mentions the findings of the Millennium Ecosystem Assessment, a body working under the auspices of the U.N. One of its surprisingstatements, even to me, is that carbon dioxide in the atmosphere is now reducing because of the reforestation which has taken place in the "rich world" during the last 50 years; i.e. in the atmosphere there is now a net sink of carbon dioxide.

We need to look for a moment at the atmosphere and to note that water vapour is also extremely important in heat retention and reflection and the fuel cell will increase its presence. Clouds cover half the world's surface and they reflect or help retain heat according to their structure, but again we have little understanding of their contribution. Dust nuclei allows cloud increase and reflectivity increase. Again it has been suggested that spraying salt so that it is taken up by clouds will whiten them and reflect more of the sun's rays. (Sunday Times 15/8/05). The sulphur gases perform a like function, and here we have opposing qualities for our claimed needs. Yes they cause acid rain, but they also reduce the absorption of heat from the sun and the methane level of production in the soil. Acid rain has reduced methane to concentrations below pre-industrialisation levels. That reaction reduces global warming by 1/5th. (Times 3/8/04). If the temperature rises, so also will water vapour concentrations. They could better retain what comes through but also reflect more away from earth. This explains why models assigned to horrify the world need the largest pinch of salt. There is no need to re-enter Eden or the Dark Ages, there are countervailing tendencies appearing ever more clearly .They do allow the prospect of development without disaster. It reveals the Luddite in every "Green".

Amid the shouting, let us repeat that the seas are full of water, not pollutants, the air is full of oxygen and nitrogen, and so it is in the earth. Even much decried pollutants, are such only in excess, otherwise we cope well.

With the urbanisation of Europe extant, there is still more than 48 million square kms (i.e. 1/3) of wilderness and we have not included areas such as Greenland. It was not man who restored the Mediterranean area from desert, but the waters of the Atlantic surging through the straits of Gibraltar. The abnormal lack of rain in the U.K. in 1990 and 1991, signalled to some the onset of the "Greenhouse effect" but 1992 gave another picture. 1995 and 1996 were very dry summers and 1997 and 1998 gave well above average rainfall. The year 2000 was the tenth warmest since 1659, and the 6th wettest since 1725. It is just that we take little note of the pressure by buildings especially in flood plains. They enlarge the misery and by placing obstacles to the flow of water enlarges the area for dismay. One million residences, 83,000 commercial properties and agricultural land are involved worth £140 billion are at risk and Prescott seems intent on enlarging the area of risk (Econ. 20/12/03). If the doomsters had done some homework, they would have seen the information indicated in a letter from Burt and Shahgedanova, that there have been repeated droughts in the last 225 years, beyond which they had no records. The earth is not uniformly a hospitable place even with man's ingenuity!

There has been such a 4 year period to 1903 and to 1893, and probably even worse droughts in extent from 1799 to 1806 and 1782 to 1789. According to a Panorama television program of 12/5/97 it was just as bad in 1859. In 2001 there occurred the highest rainfall in the U.K. since Tudor times, but we just do not conclude that there was as much production of greenhouse gases then as now. In 1606, 2,000 died as a result of flooding in the Thames area. In 1703 a storm destroyed the port of Bristol and caused major damage in S.E. England. Countless homes were destroyed with citizens dying in thousands from damage caused by winds with a velocity of 170m.p.h. In that same storm 700 boats were destroyed in the Thames Estuary and many thousands of sailors and soldiers died when 4 warships capsized on the Goodwin sands as well as Eddystone Lighthouse. In 1824 wrecked vessels littered Southern England, and Chesil village was destroyed. In 1861 over 100 ships were destroyed and 1,000 died.

In the period of 1876-9 and 1896-1900 El Nino caused "overheating" in India, North China and Brazil and some 30 million are said to have died from the consequent draught and famine. The links are however far more tenuous than indicated by the authors.

On the other side we have it that the floodings of the Ouse prevalent in 2000, only exceeded that of 1625, does this mean that man's contribution in either previous period was the same? As to hurricanes in 1703 there was a major occurrence in 1703 across southern England. The floods and the droughts are not unknown over historical periods and indeed there are some 30 tornadoes every year in the U.K. One in 1913 killed 6 people in Wales. !947 was the coldest and wettest for some 200 years. In 1770 the extent of flooding was beyond that of 2000 and there were equal disasters in 1947 and 1951, and again 1903 was as wet. Not all these can really be placed at the door of man-engineered carbon dioxide. There is also newspaper talk that for many years pleas to raise the funds for water protection by raising funds from £200 million to a necessary £400 million have gone unheeded, and that such a minor amount would have greatly diminished the problem.

First we were to accept that 1994, 1995,1996, betokened water shortage through the Greenhouse Effect, after that year it was responsible for disastrous rain and wind. The phenomenal rainfall in the U.K. in 2,000-2,001 50% above normal is comparable to the Tudor period, a time of no cars and no industrial production of any moment. Such is not the only recent attempt to assess forthcoming disasters. Many scientists and others have explained the Bangladeshi water disasters by blaming the poor Nepalese mountain farmers for cutting down the forest. The forest trees hold water like a sponge and release it slowly. Without the tree to hold back the monsoon rain, disaster was to be expected. Against this conventional view, indicated in " The Economist" of 15/10/05, there is the fact that while the forest could be active in periods of light rain, where major floods are involved, the ground soon becomes water logged and cannot function to hold areas already swamped. They therefore contribute little to holding back the flood waters. The frame accords with present thinking, but present facts also point to another scenario based on another set of causes. According to an article by Cohen in the "Independent" and Prof. Ives of Colorado, the data is inaccurate because the cutting down of the trees in the relevant area would have little noticeable consequence. Worse still is the finding by Thompson that the first conclusions on woodland loss relate to calculations sources, some being 67 times that of others. Worse again is the estimate indicating that over the last 150 years forest land in Nepal may actually have increased.

On such bases the worst of possible scenarios is presented as inevitable oncoming fact by the media. Worse can follow when such are accepted as a guide by politicians.

The famine in Pakistan has been ascribed to the Greenhouse effect (Independent 28/4/2000) yet we have Prof. Slingo at Reading University pointing out that as measured, conditions have not changed there for 100 year. He adds that there is no shortage of water in that region but just another case of bad management and poor distribution of basic necessities.

There is little doubt that the Green movement has brought to the forefront of our thinking, problems which require consideration, and their arguments should not be dismissed out of hand. Even to the extent that their analyses were partly justified, we note the nostalgic nature of their solutions which would require reversal of developments within society. Many among them are concerned to retrieve that past by reductions in population, and the diminution of our usage of energy and chemicals and not just by conservation. Their desire is to return the forest to the insects etc. We recognise that we need their active intervention for flowering, decomposing dead plants, dung and, fertilisation, but in effect the Green movement would have the soil worked so that our ability to feed the world population is reduced. It is therefore no answer to the needs of the underdog. The "Greens" speak the thoughts of Malthus and paint pictures of the new perfect savage, his tribe to be the middle class, and his life between visits to the health food shops.

We are repeatedly advised that since the turn of the century over half the tropical forest has been lost, but there have been recent surveys, less interesting to the media, from N.A.S.A. finding that half this loss has been "re-found" and statements that previous data was inaccurate. Latest figures indicate a loss of perhaps 20%.

Man Remakes The Earth

It has previously been accepted that a loss of 90% of forest would lead to a 50% loss in biodiversity. Recent facts (Economist 4/8/01) suggests 10%. Certainly a loss, but we are not God, and for the author the human race has absolute priority. Only in so far as the other kingdoms of the living play a desirable part in this scheme, albeit in the widest sense, would he make allowances for them in the scheme of existence.

The assessment of information and reasoning may not be as one sided as given by the Greenies, and it becomes obvious that conclusions spewed out today by them cannot be relied upon without confirmation, and who is to be the confirmer! Since the 25 years from 1975 there is evidence of a rise of 0.5 degrees centigrade in the temperature of the earth and this development away from previous smaller movements will probably give rise to more unusual weather. It could be warmer in the North but if the gulf stream moves it could be colder in the areas it serves. It could be drier in the South and some states will gain and others lose, but no state will wish to be a loser in these stakes. There is no evidence as yet that betokens a world food problem, but rising waters could imperil Bangladesh and low lying Cities. If Holland can cope so also could others, who have the resources.

Undoubtedly the problems do need serious address, since they pose possibilities of the greatest of disasters. But their irresponsible sightings of a wolf in every shadow including that thrown by imagination, lowers objectivity just when it could become imperative to see his emergence into life. Too many faces of doom may render him unrecognisable were he to arrive in our midst. Meanwhile, every year an area the size of Wales is destined to become desert, and double this area may become deforested and a play to erosion. The I.M.F. and world banks in their activities have encouraged the process by their ignorance, and their concern to organise countries to pay off foreign debts, especially by encouraging cash crops. Only the intermediaries on both sides grow rich.

There are some 50,000 dams built or building in 2000A.D., feeding the needs of one third of states with one half their needs for electricity (Econ. 18/11/2000). For them forests have been cleared, masses of humanity displaced and staples such as fish much diminished. China too has forced the relocation sometimes using force and offering meagre settlements. The Aswan displaced some 90,000, others 40,000 but the largest of all is taking place in China where displacement is set to remove more than a million human beings.

Now a startling note has arisen that because they also act as repositories of decaying plants they may also be making a significant contribution to the production of greenhouse gases.

Prestigious dams and other constructions have been the development a la mode and now as in Egypt are extending the areas for the spread of malaria and bilharzia. Fertile earth from Ethiopia no longer spreads along the Nile but silts the Aswan Dam, and the river salts, unable to reach the sea settle, cause deterioration in the hitherto fertile Nile delta. Worldwide (Guardian 21/5/04), some 35% of irrigated land is already too salty for many products and there is a continuing loss of some 10 million hectares per year of soil in this way. Along comes a GM tomato deriving the relevant gene from the weed thale cress and itself as a plant which is able to cope with very high levels of salt in the soil. Introducing the Futura gene has also helped rice production ,alfalfa, and corn. It seems now possible to develop plants which can survive at a salt level four times that of their progenitor. It may even be possible to irrigate with salt water and the Futura gene occurs within plant life.

Even the favoured Eucalyptus is now found to be inadequate for arresting soil erosion. Hardly half of the money spent has been fruitful in any positive way, and of the used half probably 50 % has found fertile soil only in the pockets of intermediaries. Dams also offer a potential military advantage over those occupying the lower reaches by retaining control of the river flow.

Dams do not just create electricity, they create problems for the local populations who are moved on elsewhere with minimal consideration of their needs and often less than that being just pushed out. However hydropower producing 1/5th of world electricity and saves 22 billion tons of oil cannot just be dismissed.

Dams do organise water resources and give the benefits of irrigation as against flood. They also affect the environment, the river life and (Econ. 19/4/97) displace 4 million human beings a year. Those who relied on the Nile flood bringing its silt to maintain the fertility of their soil can do so no longer, and most of those affected in this way will have little use for the electricity generated. The Dan Korigi and Bakalori dams have reduced the production of cereal and fish for the peasants below these dams. The benefit is for the arising elect seeking industrialisation.

One needs a holist view to see sheep on the Cotswolds as preventing the spread of hawthorn, when as with other animals and insects their enormous increase in numbers are the work of man.

Man Remakes The Earth

Perhaps the Green movement have a preference for the return of the wolf, the brown bear and sabre toothed tiger. We find the position where the preservation of 5,000 tigers in India is deemed to justify the destruction of 1 million livelihoods of unblessed humanity, a crime against humanity. (BBC 2 Scare Stories 18/12/97). Even the humble sheep are said to be over-cropping the grass of Cumbria and severe reductions are required in their number for balance.

The task for humanity is to strike the best balance for human life.

To return to the forest, closer examination tends to reveal that there is no Green intention to return to the previous golden era wrapped in the primitive world of gathering, hunting and shifting cultivation. Indeed there is no place assured of such a description, today. Mabberley brings to our attention that in areas of Kampuchea the oldest forests are less than 600 years of age and that many forests now considered primary in the Guatemalan forest have archaeological remains which inform otherwise. In Mexico and elsewhere in Central America and West Africa there is the same message for those who can read. It is estimated that since 1825 the forest of Venezuela increased in area from 21% of its original range to 45% in 1950 and we are not here concerned at the correctness of levels achieved for our good but just to stress that the direction is not always one way, and as indicated we need to consider carefully how far later additions should be represented as primitive forest. Again the Western World has included in its liturgy of hell to come, the dire results of the destruction of the rain forest in Brazil. Then out pops the "Economist" of 21/3/98 telling that in 1998 85% of the Amazon remains. But the stories of woe make better copy. In 2000 (Econ. 29/4/2000) Leach and Fairhead correct existing figures for the Ivory Coast of 1900, halving them, thus reducing the rate of forest loss by 40%. An important difference but even that loss should not be dismissed lightly. They, like Maberley suggest that considerable portions of forest in Guinea, Sierra Leone were farmland and savannah before that area was denuded by slavery.

Different computer programs within accepted parameters show an enormous variation in results and this should be no surprise. One has a rise between 4 C and 8, foretelling hot desert throughout mid-west America, another offers for the same period a rise of 2 C, with more rain, longer growing seasons and bumper crops. Recently it has been noted that in the last 5 years glaciers in Norway have grown more than in the last 40 years, Does this arise from a Greenhouse effect or is it perhaps a movement of the Gulf Stream giving vast effects, or both? There is indeed a wilderness of ignorance over the weighting of various contributions and even the contribution of the cirrus clouds remains clouded.

In January 2005 A.D. we have had a B.B.C. programme telling us that but for the effect of greenhouse dimming by pollutants the temperature rises would be twice as large and that our attempts at removing pollution may also spell danger. And again a Cambridge University Unit has released preliminary data in early 2005 from an exercise using the "largest" computer agglomeration on record, affirming the original forecasts may all be wrong and the new starting point is an 11 degrees centigrade rise. A preliminary statement from The Scripps Institution of Oceanography as reported to the February 2005 conference of The American Association for the Advancement of Science is that a marine temperature rise of 0.5% over 40 years as noted by them could only arise if we include as an important factor, gas emissions arising from fuel consumption. The last three asseverations have to work their way through rigorous examination before we can accept to any extent their findings. We need care in assessing the computer models chosen, the relevance of the influences chosen and omitted, as well as the weightings given these influences. However the present imprint of the last contribution on my views makes for a chink that allows us to look at a tiger's face where we have seen largely sheep, but we still do not know the actual contribution.

We are overwhelmed with doomsday forecasts and it is becoming ever more difficult to separate the wolf from among those who howl like wolves.

Reason predicts that carbon dioxide, water and methane must play a role and the possible outcome is too important to neglect just because of our inability at this stage to make reliable predictions. It is necessary to improve that reliability by continuing evaluation. The real wolf is a possible redistribution which could alter the fertile areas of the world, away from the West, and there's the rub for the powers that be! These considerations can no longer be left to their anarchic, self-interested decisions. The problem can no longer be treated effectively in terms of State agreements.

We hear already the middle class cries that the explosion of interest in walking holidays and hobbies is destroying their areas of interest. Goa is for them now demoted from a paradise to a mere tourist centre, where the "uncouth" can also take part. The farms, land and their hedge constructions, of so much interest to them are not of ancient origin, not of today or yesteryear, but they are of but yesterminute in man's history. In England they stem from the post-enclosure period.

We must also imagine the impact of the cutting down of trees not for just lumber but for the valuable pesticidal and medicinal properties possibly to be found and whether such an approach would not still lead to ends abhorred by the Greens.

We should say thank you to the synthetic drug company. They could have the same effect as turning a beautiful lonely seaside or countryside enjoyable in its isolation, into a package holiday area or holiday camp.

Green programs include renewable cropping for energy as well as cries for the use of tide and wind etc as resources for energy, but pay little heed to the consequences that may flow from such sources in side effects and after effects. The movement is in essence, urban, middle class and little concerned with the basic needs of society, nor are they in spite of their offerings, concerned with the development of science in other than an inimical sense. We are no longer moved in economic matters by religion and therefore a new acceptable mantle is displayed for our mortification, which entails a sophisticated return to the worship of nature.

Their main thrust is in the area of pollution by chemicals and energy without understanding the nature of their contribution. But they have sparked the alerts.

In the long ago, past land masses of the world drifted apart, land rising and falling in many areas and the changing climatic conditions will have had their will over the long and short terms, but we are unable to order such movements or their consequences.

When we acknowledge that the movements under the sea bed have raised parts of Alaska some 12 feet and lowered it in others some 2 feet over some hundreds of years, we should also acknowledge that there are limits to man's ability to move the Earth even though they should certainly not be omitted from our purview. Especially since we cannot affect the other sources of our problem, it does maximise the importance of relating correctly the level of man's contribution to the Greenhouse effect.

The present schools of thought relate the extinction of the great reptiles to catastrophe, perhaps a vast comet impact, a volcanic or earthquake eruption causing changes similar to that already offered in various nuclear permanent winter scenarios. Apart from possible developments in anti-ballistic missiles to deflect a comet of this size, little can be envisaged by man to evade such and they have occurred!

Volcanic eruptions may have played a part in cooling the world each contributing to emissions of millions of tons of sulphur dioxide etc. There was Tambara in 1815, producing 40 cubic kms of ash ,leading to crop failures, and famine in much of the World, with an estimated loss of 200,000 lives. That eruption, ten times the size of Krakatoa extinguished the sun in Europe for a whole summer. Yet Krakatoa was not small in impact, since a whole island was vaporised, costing tens of thousands in lives. The 17th century records that a similar event in New Guinea gave seasons of darkness in Europe. The Santa Arene eruption could have been the means of destroying the Minoan civilisation and even contributing to the woes of the biblical Pharaoh. Recent eruptions back to Mt Aetna are documented but less is known of the Mount Hekla event of about 1150 B.C.. In Scotland, the inhabitants were decimated and the areas denuded of vegetation as were the areas about northern Scandinavia and Ireland. Sulphur acids were abundantly distributed contributing for at least 20 years to a wet and cold climate. The Caldera gave 5 years of winter. In correct perspective the recent smaller eruption in Columbia was equivalent to some 500 Hiroshimas and St.Helena denuded an area of 270 square miles. Koba in Sumatra destroyed the nearby town and the sulphur dioxide encircled the world for two years. The July 1997 eruptions at Montserrat and Popocatepetl seem small by comparison even compared to that which took place in N.W. Pakistan which seems to have killed some 75,000 in 2005. There is also the Indian Ocean tsunami which as a result of tectonic movement destroyed some 200,000 lives. We seem to have to suffer some 20 active volcanoes and some 50 eruptions yearly. 65 million years ago the reptile age ended suddenly. Again when one looks back 440 million years 60% of marine life ended quickly and 250 million years in the Permian period when 2/3 of reptile and amphibian life, 1/3 insect life and 90% of sea dwellers were also quickly exterminated one can see that man-made disasters projected or otherwise are may have no great significance in the SCHEME OF THINGS. Nature is profligate with life and our contribution puny in comparison and the insect and animal counts by men have little basis in fact and have some affinity with myth. The U.N. "State and Marine Environment" issue of 1990 informs that volcanic eruptions on the sea-beds give heavier pollution than man in respect of heavy metals and radio-activity. Indeed they feed a different possibly earlier life form based on the sulphur cycle. All that man can concern himself with are those stemming from himself, but the present projected variation is well inside the known variation in carbon dioxide since life began and continues.

War, with the present potential of nuclear energy, has now stepped up to prime rank in destructive potential despite international laws and so-called embargoes and is a major threat to existence. Weapons could destroy nuclear generators, unleashing their potential for further extending the damage caused. Apart from such weapons, war destroys the means for sustenance. For years Ethiopia was in the grip of war and famine, yet today in 1997 it has a cereal surplus.

In our previous paragraph, we have indicated that the good earth in a malevolent phase dwarfs all our technical means for destruction and we have yet to discuss its role with respect to radio-activity. The multiplicity of nuclear powers capable of being directed at the centres of technology could become nightmare enough. The free market for capitalist weaponry is world-wide ensuring that no measures can limit their proliferation. Man does need to decide for his advantage and by any accepted ethical code whether any level of displacement of the insect world and animal and vegetation by himself is justifiable. An Economist contribution confirms that bees are very important to us since they play a major part in fertilising about $20 billions of U.S. crops , including pears, citrus fruits and apples.

To the extent that our forebodings are beyond securing within the present structure of society this gives no entitlement to throw up one's hand in a gesture of despair, and war at least, is a social problem and within social solution.

The mode of organisation and the basis of our subsistence within the present phase of our social structure is the continuous, assured supply of energy to feed production. Energy is synthesised by plants reacting in sunlight with carbon dioxide and is the source of energy today sustaining life for the animal kingdom. It has also supplied the stores of fossil energy laid down in the past. There is also solar power, hydropower, tidal power, wind power, geo-thermal and nuclear power. To this one should add the relatively unrecognised and gigantic potential of available shale oils and tar sands, largely untouched. In Canada alone there are 1.5 trillion barrels equivalent and some 300 billion (about half the known reserves in the Middle East), available with present state technology at 18 Canadian dollars per barrel and with hopes of a reduction to 12. And this oil is sulphur free and thus would carry a premium rating. (Times 5/3/05)

At a certain level which may be critical, the side effects arising from the use of these may engender deleterious defects needing to be weighed against the benefits offered and the potential of other sources which will also have drawbacks. Were the dangers inherent close to causing disaster, then it would be reasonable to forgo as necessary, such use whatever its qualities. The means to overcome these are often available or foreseeable but can fade into a no-go area in the face of short term economical considerations. The problem is to keep deliberations quite open so that pollutants unrecognised heretofore should be allowed to emerge.

The solar thermal power intercepted by the earth is 17.7×10 to the sixteenth power in watts, about 100,000 times the world electricity supplies. Some 20% of the solar radiation received contributes to keeping our planet warm and in our present stage, less than 1% is utilised in the operation of photo-synthesis for life. Prof. Cox (Indep. 17/4/97) tells of 7 thousand billion tonnes in use from fossil sources. In the grandest of expansions in their use by the developed and developing nations, a 40% increase would still be a reasonable foreseeable ceiling for mankind. There are major savings available from man's actions in improving insulation, dwelling designs or developing major efficiencies in transformations to usable energy. Since feeding our cereal crops only costs 50% of the fuel present, we have a ready means of reducing our energy usage by extending the area of tillage, but today's grower obtains some 3 times the recompense by selling it as food, not fuel. In such use there would in effect not be an increased production of carbon dioxide but effectively its recycling.

Were we able to consider the replacement of fossil fuel by other means now available, then fuel cost might well increase two-fold, an increase well within the yearly increase in World G.D.P. Oncost is by custom treated not as an arithmetical addition but on a percentage basis. This method of accountancy means that raw materials and energy used at the initiation of a process give rise to increased end charges and exaggerates the real cost. However if society can afford the "millennium computer bug " variously assessed at between $2 trillion and $100 billion (Econ. 4/10/97), a new look at fuel sources should not be totally out of place.

Projected figures for the resultant use of fossil fuels point to the increased concentration of deleterious gases such as carbon dioxide which by their nature enhance the retention of solar radiation and such a development is considered dangerous for our viability by some scientists and increasingly, and more importantly, by politicians marketing their wares. The latest figures of NASA (Ind. 9/12/98) alleviates the doom picture somewhat informing us that plants in the northern hemisphere absorb about 1/3rd of the carbon dioxide produced by fossil fuels, and probably as a result plant growth increased in that region by some 10% between 1981 and 1991. However there is little doubt that fossil fuels, very much a wasted asset, could be better used for their chemical potential.

Little if any attention has been displayed in the direction of increasing our ability to capture a greater part of the enormous renewable energy fund available to us seemingly nearly free. We look however to the past to retard, diminish or prevent usages, rather as one who prefers to work against nature than with it. There is for example the process of photosynthesis itself where all plants are lumped together, but it would indeed be the greatest exception were these not of varied efficiency in this conversion . We therefore have the means worthy of pursuit of say doubling plant efficiency for our food or fuel by choosing or developing plants superior in this respect. With our increasing ability to apply our knowledge of genes we have no need to fear even lack of variation in this respect.

Man Remakes The Earth

There have been explosions of growth in human numbers, cattle and agriculture and a decimation of the vast forest areas of Europe including England. Great reductions have occurred in the rainforests areas of the north as in Canada and the tropics around the world but especially in South America, regions bordering on India and parts of Africa. All these alterations have left man dominant and ostensibly with an improved inheritance. Nobody seems concerned with a need to restore to their full extent, or indeed to any extent the forest and the forest animals of Europe and the reason is not just due to the superior land produced therein for agricultural purposes. The forest area cleared in Europe since 1000 B.C. equals in extent the Amazon and we mourn none of its destroyed species.

The problem of the rain forest, certainly the tropical forests, is that much of that land as at present cultivated is inadequate for long term use and that the progression strikes the West as adverse to its long term comfort. In this scenario the West is not concerned with immediate needs and still hopes at small price to appease our sense of guilt towards others. It could even result in a gain through an attempt to reverse the clock, and "save" the quality of the air WE, WESTERNERS, breathe. The data fed seldom includes mention that this destruction is immediately replaced by grass and cereals which also contribute to the carbon cycle. UNEF have now discovered that tropical grasslands account for 25% of all photosynthesis and on this basis take up three times the level previously ascribed them and burning grassland gives rise to only 1/10th the carbon arising from forest burning of the same area. The problem is burning not replacement and of improved use of either residue. The tropical forest in one single year will ingest as much carbon dioxide as the tropical grasslands. A mature beech tree produces as much oxygen as 600 sq. ft. of lawn (Indep. 21/10/97). When one looks at pictures of the rain forest one is struck by the paucity of greenery below the top levels and which suggest that in this respect the cycle might be better advantaged by broad lanes of open grass or cereal areas which would allow the growth of tree foliage at the lower levels and the whole displacement produce the means of reducing carbon dioxide in the atmosphere. There is now also the possibility of trees modified using G.M. techniques etc able to give a better recovery of carbon dioxide from the air, offering more desirable qualities, and even faster growth.

Husbanding world resources, depleting none, requires little improvement in efficiency or energy savings and much is already within our domain of knowledge. Frynas' letter on Nigeria ("Independent" of 16/5/97) states that 2,000 million cubic feet of gas are flared each day sufficient to meet the total needs of Holland and now under pressure, Shell propose a possible interest in this source of energy but only for the distant future. More recently (Indep. 17/10/97) we hear that Shell have expressed an interest in renewable trees regularly cropped with the new grown taking up an equivalent amount of carbon dioxide to that burnt, but one would need to know more about the relation of tree to site before this can become a positive contribution.

Why then all the convulsive twists and turns felt as necessary, not only necessary, but fast becoming an integral part of the body politic? It is not just concern for the number of species supported in these forests and to be found nowhere else. A market orientated society precludes the possibility of maintaining or advancing species even man where it is unprofitable. Thus the bemoaned destructions will continue even where the data disguises its nature and its extent and there remains legitimate fears of their destruction beyond recall. Instead of whales killed for the market, they are killed for "research".

Water resources, possible movements in fertile belts away from the West, levels of the ocean exercises us even more. Without the presence of Man, in the Cretaceous period, there were tropical plants in the temperate zone and most of the polar regions were free of ice, making for high water levels. We are made well aware that our water usage is abominably wasteful in application, more than half of that applied just evaporating, but there is need to examine more carefully what would result in the wider cycles of water and carbon where it is changed. It is claimed that the tapping of ground water in Bangkok has caused a general subsidence in the area of a metre over 30 years. Unused evaporated water also plays its part. That superfluity may feed other important but little understood cycles and cloud formation. Taking too much could be counter-productive.

Why is so little achieved? As well to ask why the clear message as to smoking has needed so many years for general acceptance, why when economics is so simple, so many convolutions are deemed necessary to hide the intent and why the simple message of socialism is so far shrouded by nearly impenetrable obfuscation in the way life is organised for us. Largely the main action arises from the division of wealth within society by our betters. All are brothers, occupying no more than six feet in death and the solution to world-wide problems is not to export the problems. It is bound up with the world framework allowing adequate and available technology and resources to be freed for use. In too many necessary directions for an efficient world operation, the state organisation is past its time. It can no longer protect its minions, from the ravages of a world tearing itself apart and has too little control over the global use of arms, energy or trade.

The growing levels and nature of concentrations of carbon dioxide, ozone, methane, sulphur gases, nitrogen oxides and last but not least the C.F.C.'s are exciting concern, even panic, in sectors of society. Little account is taken that approximately 1/2 of our nitrogen oxides arises from putrefaction and residues of organic matter and in large measure they arise from the tropical forests. Again termites in that region and elsewhere produce enormous quantities of methane and our cattle produces 1/3rd. The Wetlands of Canada have doubled their production of this gas over 50 years, increasing by about 1% p.a.. When there is an attempt to reduce the marshland of Iraq, it is treated as purely a question of political advantage by all sides. Indeed ,are we to limit rice as a food because of its high release of methane in its growth process. The Indep. of 17/5/93 tells of a widely spread microbe capable of converting methane to methyl alcohol. Certainly their development can only be treated on a one world basis. The major problems highlighted are the greenhouse effect and the ozone "hole". The first takes into account that in various degrees, increases in carbon dioxide arising from fossil fuels and the destruction of forest land and production of methane help retain a greater share of the heat received from the sun. Such a development, might move rain in a direction deleterious for some, or all, even advantage some. While there is a possibility that world wide, such a movement could perhaps have an overall advantage, the Major Powers have no wish to risk diminishing their share of the cake. They would not be gratified by a blooming vast N.Africa at the expense of a Kent desert. They would prefer status quo and retaining the Best as in other fields, but as in other fields are yet the main agents causing these changes. Schumpeter, not a Marxist, recognised that new technology leads to the destruction of old social relationships, even undermines the world status quo.

The other fear is that even minor degrees of warming may cause melting of considerable dimensions at the polar snow caps and that the rising waters could adversely affect valuable lowlands especially the great centres around ports which are so important in maintaining the fabric of western and other societies. But where this led to the higher lands becoming more fertile and other lands finding benefit therefrom these could be in competition with the West. There may be advantages such as the opening of the North West passage for the first time saving 5,000 miles between Europe and China. Cod and whiting are diminishing in our waters but anchovy and red mullet show an increased presence. Crustacea are now becoming a presence in the Canadian zones where cod formerly reigned and as we learn more about the habits of the cod ,their return to these regions in quantity is highly possible , provided we are prepared to maintain rather than destroy. In the last 10 years plankton has increased manifold busily benefiting from the carbon dioxide levels, and sea life has doubled. There is therefore more base food in the seas possibly spurred on by our depredations among the species higher in the food chain. Perhaps we need to revise our eating habits. Indeed Phytoplankton according to Scientific American August 2002 apparently absorbs an amount of carbon dioxide equal to that absorbed on the land mass and is important in retaining carbon dioxide in the waters. They also have preliminary findings where addition of iron can lead to their proliferation, thus affording another possible instrument to control the atmosphere, but it is not easy to see why we are not opposed to anything, which could be a factor in preventing another ice-age.

Meanwhile, we have D. Vaughan of the Scott Polar Research Institute at Cambridge informing us in 2001 that although the Antarctic Peninsular has warmed since 1945 it forms only a small part of Antarctica and that little shrinkage of the ice sheet is to be seen. Again the Sunday Times reports that in the period 1975-2000, weather satellites and balloons give no clear sign of warming. Hansen, a major contributor in this field avers that we could comfortably accommodate a rise of 1 C, which he reckons is the foreseeable rise over 50 years and there is an IPCC report stating that there has been cooling in the period 1940-1975. It seems that although sea levels may rise from the melting of non-polar glaciers, Antarctica makes no contribution to sea levels. Indeed as Antarctica warms there will be more moisture falling as snow in that region and held in this form. He suggests that where there are retreats on the peninsular this is due to the happenings at the end of the last glacial period, some 10,000 years ago.

Although nuclear energy does not give rise to these problems it gives fright for the possibility of a nuclear winter which could end society as we know it, even end life itself, but were this not the scenario to be played out we still have the spectre of spent, radio-active material which we do not know how to safely withdraw from society. It is true that these residues are small, but it could mean a large problem for the future. As will be shown later the dangers and the consequent costs could have been avoided, had one not been misled by the siren voice of market forces and state politics.

Were all this to be an accepted interpretation of events one would still be struck by the backward look of the backwoodsmen. Here is a source of energy and all that can be seen is the flame hiding the energy of which it partakes and the continuing wish to maintain the world status quo in all its manifestations. To such the invention of fire would have been anathema. Yet the world climate like all else in our lives is ever changing. We had a lesser Ice Age 5000 years ago, in the Middle Ages we had a mini-Ice Age and according to some experts we are on our way to another in some thousands of years to come. Yet the message given out is that no effort is desirable to stall this adverse phenomenon even if we could at least lessen its impact. Surely no one considers The Ice Age as a time of great advancement and it is not rational to see any better promise in a future Ice Age. Yet few view the possible greenhouse effect as offering a potential advantage in off-setting the onset of the next Ice Age. Nobody thinks that we should work with nature, as part of nature, to advantage us. Our anti-scientific society even uses selective scientific data to support Luddite activity.

As totally isolated happenings the conclusions and their likely consequences have a reasonable ring to them and they do paint pictures of despair, especially for the longer term. However social problems are seldom simple and one has to deal with the complication ,other "outside" factors bring, especially when they counteract each other or are of differing moment. Unfortunately society has not yet accepted that not all problems have a simple answer and our leaders do not accept that we can all understand quite complicated solutions for complicated problems when they are put to us in plain language.

During the last 2 million years (Econ. 5/12/98) glaciers have been present in the tropics of today. They will return in ecological time unless we can contrive otherwise. We must not exaggerate the importance of man's contribution to the greenhouse effect but it does give us a glimpse as to a means of avoiding the worst effect of an ICE AGE or HOTHOUSE AGE.

Let us therefore reiterate that 20,000 years ago, one third of the earth's surface was covered with ice to a thickness of ten to fifteen feet much in places where we now live comfortably, and the average temperature of the earth was only some 5 C below present levels. Let us be concerned that we have had some ten Ice Ages in the last million years and the earth has spent more time in the cold than we have. Blindly accepting such an event we will be entering cooler and cooler conditions leading to the next ice-age. Even were the time span for entry 10,000 years, mankind could be suffering consequences prior to its onset. It will be cold and present crops might cease to germinate. Herein lies the important potential development of gene understanding so that new problems of a developing Nature can be ameliorated, even avoided. We must not be so overwhelmed and blinded by the concepts of the status quo or back to Eden, not to realise that were such a period to occur, man would need all his efforts and knowledge to maintain a society comparable with today. Some of us may not wish to bestow that heritage for those who come after us. To maintain present viability levels, we will need all the help we can get from our present day knowledge of gene structure, and the greenhouse effect. Of course any attempt at alteration has its dangers but so also has the do-nothing attitude prevalent today, coupled with an existing back-to-Methuselah movement. If we attend to it, we may just be able to use global warming to help control the environment.

There should be enlightened concern with the amounts of carbon dioxide entering the atmosphere, because there are dangers arising from lack of control, but it has already been confirmed that even without the benefit of higher temperatures premised by the greenhouse effect many plants grow better and faster at higher levels of carbon dioxide. Preliminary findings however indicate that those we classify as "weeds" seem to do better than man-developed plants, but a more adequate knowledge of the D.N.A. structure could help here. There is also an assessment by Professor Teeri of the University of Michigan, that doubling the concentration of carbon dioxide would stimulate tree growth by up to 20%. In other words there is always a tendency towards equilibrium conditions, but we are never able to say when a new unwelcome equilibrium level may result and there is the rub. Then again there are claims that increases in carbon dioxide levels could decrease the stomata content of plants in the longer term as the alternative answer to maintain equilibrium conditions.

There is a peculiar hole in greenhouse calculation which cannot explain the "disappearance" of some 50% of carbon dioxide from these calculations. It may be that some escapes the earth's atmosphere but provisionally the only reasonable sink is the sea. A growth in algae population has been noted, again with doom-laden ascriptions but it is feasible that much is absorbed and converted to carbonate settling in the sea. There is also the formation of carbonate from the interaction of the dissolved gas with calcium silicate on the floors of the oceans releasing silica. By such means carbon dioxide would be safely removed from action. Higher temperatures might only mean an increase in some species and more carbonate formation. Absorption into the sea or its presence anywhere may offer advantages and disadvantages to species, affecting their distribution and extent but it would be wiser if we looked more closely at such huge unexplained probably very important amounts. Even a Minister might see a need to explain the disappearance of half the sums he expended. It may be that certain equilibrium forces are brought into play to balance out or ameliorate the impact of these amounts. Certainly we should not accept a view that it must end well, but equally the opposite view should not be chosen without some greater effort to elucidate the situation which could be very important to our existence. Were the loss by diffusion greater than expected then other considerations might surface.

Above all when one is told that one quarter of animals, insects and plants may disappear in the next few decades as a result of man's destruction of their habitat, we should not play Noah and we should recognise that it is not for us, a must, to preserve the status quo in species and their numbers. Some species have grown vastly as man has extended his development of food and this is especially the case with insects, rats and domestic animals.

After all, our social and economic practice is to extol the virtues of the free market where the weaker among HUMANS are pushed to the wall or at best are held in abject poverty. Certainly one may lose the contributions that a few insects or plants might produce, and certainly one can lose the benefit of their gene banks, but with our new light into the structure of D.N.A. one may be able to create new and better forms backed by knowledge.

We will never discover in any field of enquiry all that is to be known, since that was not the original direction for our understanding. Destroyed forests will take generation on generation to re-grow if ever but there are many species in England deserving of consideration. We do not read of forays into Kent to preserve the apple bank or some of the home vegetation destined for destruction and perhaps the reason is that a free trip to the Amazon conjures up a different picture.

It is perhaps pertinent to remind the reader that in the period between the last two glacial periods, temperatures began to increase without contribution from carbon dioxide and at the onset of the last Ice Age temperatures fell in spite of increased levels of that gas. The reasons are well understood in their astronomic setting. They are not paraded by the greenhouse lobby, in spite of that fact or perhaps because it may show that there are still stronger forces deciding man's destiny, than appear as bases for their long term forecasts.

In any case we have seen an affirmation stemming from Professor Beyer of the M.I.T. apportioning the importance of contributions to the greenhouse effect as 99% water vapour and cloud induced and 1% to carbon dioxide. Prof. Lindzen of M.I.T. calls the latest contribution of I.P.C.C., waffle, in that the changes about us are within natural variability. The forecast they chose, involved choosing an amount of aerosol designed purely to give it the answer it wanted, and its Head of Technology agreed that some of the factors fed into the computer forecasting software was not properly understood. Lindzen's measurements show cooling in 1940-60 and no net increase since 1979. Even I.P.C.C. see no change in global food production (Indep. on Sunday 15/10/95), and from Scientific American 3/97 I.P.C.C. do not project any change in the polar ice-caps. but a change in site fertility could upset the important status quo. Again there is movement in land up and down and Stockholm has seen a drop in sea level and Honolulu a rise. Were ice caps to go then the earth below could also move up.

These contrasting assertions arise indeed from scientific data, and confirm that in these fields matters are not as clear as some schools and especially some political lobbies would have us believe. The important threshold at present is that a rise of 60 cms over the next century would cost the Dutch in dyke maintenance no more than the maintenance of their bike lanes and the latest finding (Sci. Am. 3/97) is 2mms rise p.a. over the last 40 years.

The Greens are more circumspect concerning the role of methane and treat its production as a joke about ruminants, but their level of production is quite high, 13% in Australia. The paddy rice fields are an important source. It is produced in plant ripening, in decay, and of course there is that arising from sites chosen by man for organic waste. Probably the vast swamps remain by far the major source as the plant life therein decays. As we have already noted the swamps and wet forests of Canada have doubled their methane production over the last 50 years and continue to increase at a level of approximately 1% per year. The Greens wallow in swamps but while it is true that methane is a more effective warmer than carbon dioxide concentration for concentration, it does have a shorter life.

Many will not be aware that when the car was introduced it was considered by many as a contribution to reducing the pollution in the streets caused by horse manure and urine. As with medicines we speedily dismiss the old problems as they were solved into oblivion as though they could never return.

In cars, conversion of nitrogen oxides, carbon dioxide, carbon monoxide or ozone by catalysts will not affect the greenhouse effect since their converted products will largely act in the same way. The marginal improvements are in reducing noxious carbon monoxide. The unpublicised fact is that the usage of fuel and therefore production of greenhouse gases is greater when catalytic converters are used.

With respect to temperatures some tell us that the earth has benefited by a rise in global temperature of perhaps 0.5 C during the last century, but there is no certainty that our path towards the next ice-age has not played its part in reducing it to that level.

During the Kuwait War of 1991, King Hussain of Jordan and the Greens were describing vast damage to the atmospheric scene and the waters if Hussein of Iraq released oil in Kuwait and fired it. Greenhouse warming would get out of control, life within the Indian ocean would be decimated and made unacceptable as a food source. The war over, the cries subsided in recognition of a burst balloon.

What is offered as experimental evidence is that the rise in temperature is correct within 0.3C and there are further natural variations over the relevant time interval of 0.2 to 0.3 C from natural internal earth fluctuations. There is also a variation of the same order over 11 years from sunspots.

The maritime nations have been measuring temperatures at sea for some 150 years and the more advanced nations have been making many more measurements on land for over 100 years. Between 1880 and 1940 global temperatures increased by 0.25 C, between 1940 and 1970 there was a decrease by about 0.2 C and between 1970 and 1980 there was an increase of 0.3 C. However the measurement methods of today vary from the past because of the inaccuracies of the latter and that level of inaccuracy is very difficult to assess. The Victorian ships used canvas bags and there are now seen inherent possibilities of error in their measurements. Their conditions and timing would be important. There is an important difference between night and day measurements, yet no specific control was exercised on such a point. As to land-based measurements, earlier calculations did not take adequate account of the proximity of the chosen sites to urban centres, and their creep with time towards the measurement site will have made contributions to increases in the temperatures as measured. There are also those computer models for forecasting not next week's temperature, but that applicable to a hundred years away, and whose merits need constant revision. The public acceptance that computers imply absolute accuracy assumes that they are fed absolutely accurate data, correctly weighted. What is given us, is that the temperature rise of 0.5% is accurate within plus or minus 0.3% . Our results need allowance for instruments of measurement having changed and the changes at sites for measurements. To say this after all the above, means that it is necessary to acknowledge that in procedures and therefore conclusions, we are at "baby crawl" stage and the "baby talk" arising is accorded a higher respect than is warranted. Babies make an enormous amount of noise some based on real, some on imagined fears: it is for the mature public to elicit a more mature picture.

Indeed (Sunday Times 8/6/97), Washington University stressed the need to go back to the "drawing boards" since the development of global warming is seen not to accord with all these dire predictions. There was the conclusion that the computer simulations are not adequate and must be considered as made while omitting essential but unknown data. It is a fact (Econ. 29/11/97),that no presentation at present available predicts the past with any precision, and many tests of computer projections , when applied to past events show twice the warming that has in fact occurred.

Let us be clear that the investigation of this phenomenon is of the utmost importance and that great strides have been made, but a little humility in presentation and assumption by many doom watchers would not go amiss even if grant support suffered. Unfortunately the knowledge that balanced views are not the best entry into the media world has become too important in the scientific world, the aim being totally in favour of earliest publication.

Little is known of the role of the oceans and their undercurrents, in all this, except that it seems an enormous sink for these gases. There is too little regard to the role of the oceans which may hold some 50% of the carbon dioxide produced. Plankton and other forms are able to absorb this material and form skelatae of carbonate, which are removed from the cycle by deposits in the sea, revealed when lifted from the sea level as cliffs.

The incidence of sunspots is known to be important for the ruling levels of temperature on earth and their level of contribution still an unknown, and recent data from the 1999 eclipse of the sun has indicated that the sun has become hotter. If confirmed it will make its own contribution without any assistance from mankind.

Prof. Barrett points out that a prediction of 3.5% C over the next 100 years is within the error limits of the work done and does not offer support or otherwise to positive let alone alarming conclusions.

The Ozone Layer

The highlighting of the precarious stability not only of the atmosphere but the stratosphere surrounding it, closes a time when this was also considered as set and unimportant to our lives. Radicals, even free radicals are to be found here as well as in politics and could be much more dangerous, and it may be posited that we play a dangerous game in this arena too.

Indeed, out of the blue came the sounds of new alarums, speaking of a phenomenon, now known as the Ozone Hole. It is only since 1978 that serious charting of Antarctic assets began. As a bye-product of such a survey, Farman in 1985 showed that during the 1977 -1984 period there was a regular drop of some 40% in the ozone concentration in that area every spring, and a full yearly recovery thereafter. The stratosphere concentrations of concern are of variations of the order of one part per million and detection of such a level needs sensitive instruments only just then available. It needed an unusual scientific interest to aim at all in that hitherto neglected direction. For that period it has been an Antarctic phenomenon but it seems to have widened its area and extent since that time but there was no worsening in 2000A.D. While it is easy to dismiss it as a hole that never was, ominous possibilities present themselves.

That development could pose a dangerous threat to mankind and a threat to existing life. The fact is that we are aware that ozone in the upper atmosphere exercises a protective influence on life against the baleful result of a yet higher level of U.V. than is prevalent today. The more penetrating part of this range still enters with some hindrance but the remainder has its level materially reduced by ozone absorption in the upper layers.

The villain of first choice was nitrogen oxide, already established as the inevitable chosen culprit for many ills, but was speedily rejected because of the short life envisaged in the conditions extant. Attention was then drawn to chlorine compounds as the probable chain activators. It was already known by 1974 that C.F.C.s could be active under such conditions in the stratosphere for a life of some 50 years and there were perhaps compounds such as dry cleaning chlorinated solvents also to be considered. The possibles now include the even more active bromine compounds used in fire extinguishing and methyl bromide used as a pesticide spread by air to the extent of 60,000 tons per year. Bromine seems in experiments to be about 100 times as active as chlorine compounds in destroying ozone, but does have a shorter life. Such compounds arise from ocean algae who are responsible for some 200,000 tons per year of this compound and they have been in engaged in this activity a very very long time before man. The "Greens" only tell us that algae growth is a threat to sea life and caused by man's pollution. They prefer to concentrate on the C.F.Cs because they are of man-made origin. However despite these forebodings, ozone destruction is accepted as peaking in 1994 and being on its way down. (Economist 19/4/03).

Despite this comforting message the scenario needs further review. Its importance should not be lightly dismissed and we develop the present position in that light. A loss of approximately 7.6 % seems to occur during winter in the U.K. and surrounds and seemed a norm until 1997 seemed to indicate further deterioration. It is as yet too early to place the variation within an already existing cycle or as a developing deterioration, and there remains room for concern. 1987, marked a fall in one day of as much as 10 % in the Antarctic over some 3 million square kilometres, and certainly no chemical interaction could explain that day. There are still very important unknowns here but progressive reductions in ozone concentration in various areas is on record. Reduced concentrations of 1 %, overall, has been forecast to increase cancer by more than 3% and the problem is important within this context too, since some figures now indicate reductions of 2.9%. There is also a forecast of 5% reduction by 2,000 A.D. and while this is not the end of the world, it could be the end for very many people and could affect the cycle of life.

Concentrating on C.F.C.s however, poses a problem, since these chemicals were not available 50 years ago and hardly in important use 30 years ago. They diffuse slowly and if it takes 80 years to reach the major reactivity site of concern and then have a life of 50 years they have as yet played little if any part in the present scenario unfolding above us. Whatever takes its place in this reaction, will have needed to be produced before the advent of C.F.C., and may still be in production today. Perhaps an as yet unrecognised wolf, is stalking us at this moment. A real wolf producing chemicals which are not man-made could well appear more relevant. We have also to accept that more C.F.C.s to the extent of 18 million tons are still held on earth, far more than has ever been sent into space. The good news, the result of pressures contributed in major measure by the "Greens", has been that companies are now replacing them but with more expensive if less efficient substitutes. These are claimed to contribute 1% to the greenhouse effect against a previous 14%, and play no part in the ozone reaction. Alternatives to the use in manufacture of C.F.C.s in high-tech cleaning applications are also now available.

There is data to show that under conditions where the ozone has most thinned, there is an effect on the production of plankton a basic starter material in our life cycles. It can affect photo-synthesis and affect the gene structure and these have wider implications for us than the effect on just plankton. It now seems possible that this cycle does not lead to full recovery and thinnings are beginning to appear even in the Mediterranean area. In ecological terms man has however suffered cycles much longer than can be ascertained here for a phenomenon first noted in 1985.

For plankton the concentrations of the relevant range of U.V. may result in a change in population favouring species best able to resist its effect but this basic food of our planet should remain. The picture of gene structure changes is more blurred .

Cries of doom arise over exposure to sun and to possible increases in the incidence of melanoma and blindness etc. The ozone hole today is bigger than ever noted and there is a report that the increase in loss over ten years to 1991 was four times as great at the South Pole accompanied by an ever extending area of depletion. That conclusion is still vitiated by lack of measurement prior to Farman in 1985, and he is nowhere near as downright in his conclusions. We have a NASA report telling us that ozone in the upper atmosphere has suffered a depletion of 2.9% over the last decade, sufficient to lift the problem to the danger threshold. At this stage we just do not know the answer.

With the possibility of such menace, caution needs to be the rule. Since there seems to be a yearly cyclic pattern, and there would seem to be no such cyclic factor present in the use of C.F.C. That cyclic factor must be sought in weather patterns or in the life cycles of bromine producing algae for an explanation. It is necessary to better understand the variability in the proportions of ozone, and why the variability is greatest in the Antarctic, indeed why that region seems most affected by gases which are released in the northern hemisphere. It is known that sulphur dioxide can destroy nitrogen oxides which could otherwise remove chlorine radicals from reacting with ozone or on a short life basis itself, destroy ozone. This picture of interactions is much more complex than has been indicated in the media. Especially is it complicated by events and interactions not fully understood. It is not due to the inability of the scientist to communicate, sufficient clear data is not yet available . The C.F.C. contribution may only act as a possible final straw yet to come. While the use of man-made chemicals can be stopped, those already in the upper atmosphere and from other sources are less tractable to elimination. A long wait is ahead, but not yet a wake. Meanwhile it would be good for the public to be offered a rounded appraisal and not used as a battering ram between pressure groups.

In the "Independent "of 18/9/02 there comes the informed view of Australian government scientists that the level of ozone depleting chlorine in the atmosphere has been declining since 2000 A.D. The result as seen of the original C.F.Cs having been banned in refrigerators and air conditioners after agreements reached in 1987. The prediction now is that the ozone hole will begin to diminish by 2005 A.D. and disappear by mid-century. But we are not that assured.

Thus at a slight cost in the efficiency of the replacement coolants problem may well be solved, and another green Wolf slain, this time by the positive actions of world society producing an ameliorant or cure.

Ozone concentrations were lowered considerably following the El Chinchon volcanic eruption and this will be true in others, despite high volumes of sulphur gases. The concentrations vary within seasons and with seasons, and they seem to be sensitive to our wind patterns and there seems to be recoveries.

Combustion products arising especially from road transport are considered an evil in the lower atmosphere, not only because of the production of carbon and nitrogen gases, but because of the associated release of ozone. It attacks the lining of lungs and interferes with plants growth but it should diffuse into the upper atmosphere with time, or convert to oxygen. It may just be that the evil caused by ozone is not as important in extent as has been suggested, not that we should encourage road transport for that reason. The exhaust emissions from aircraft are quantity for quantity some 30 times as great in their greenhouse effect because the release is on average at some 7 miles above the earth. They already produce 3% of the World's carbon dioxide, but they still have some way to go before they merit the disaster litany which is pinned to their shirt-tails. The nitrogen oxides also released is bad for the greenhouse effect. It would seem a requirement to devise an ozone that does not have a greenhouse effect by forming only at the "correct" site, and which excludes any site used by man.

It is known that releases from volcanic eruptions can diffuse to the outer atmosphere, apparently too, C.F.C.'s can enter this region from canisters of shaving cream and deodorant, but definitely not ozone. Even were there great differences in reactivity it would still be hard to explain. There have been countless volcanic eruptions and yet all ozone has not been destroyed. Man needs to know how ozone reaches and is established in the heavens.

The use of C.F.Cs has now with hindsight introduced a danger by its role in the upper atmosphere in taking part in a chain reaction which destroys ozone. Chain reactions under little or no control as with nuclear chain reactions spell danger. Unfortunately there will be for many years a diffusion from past emission of these C.F.C.s and things may get worse before they get better, hopefully not dangerously worse.

Replacing C.F.C.s with say vinyl chloride and other propellants for domestic, gardening and cosmetic chores, could introduce other insidious faults since in use these can enter the lungs or diffuse through the skin. Man was able to exist and engage in social activities before propelled de-odorants and if we must have them then roll-on, or solvent techniques should be the way, but for existing stocks bacteria which is known to convert methane to methyl alcohol has been found able to degrade H.C.F.C.s and C.F.C.'s.

Feeding the Hungry

Mankind uses less than 1% of the energy the earth receives from the sun. Were double that amount used, then the balance of nature is unlikely to be affected.

We till 11% of the earths surface out of a total of 24% available arable land, but of course the distribution of land and workers varies throughout the world. The U.S. a largely self-sufficient food economy, employs some 7% of the population. In the U.K. half the food is imported and the other half involves a working force of 4%. Modern agriculture produces twice as much grain that it did 40 years ago and on only 1/3rd more land. (Economist 23/8/03)

Cereals provide some half of the world's food (maize, rice, wheat) with animal products, root crops, fats and oils, fish making up another 45% . Cereal yields vary according to the state of the soil and the expertise and state of organisation of the society, showing 5 tons per hectare on some sites and as little as 1 ton in others. The state of the soil and the expertese are factors capable of much improvement throughout the world, only the geographical site is fixed.

Within the present limits, world production of cereals has reached 7,000 calories /day/person (Cottrell, Environmental Economics, 1978) which is reduced by the depredations of pests, disease and insects to 3,000 cals. This amount parallels the corresponding equivalent need for protein. There is fed to cattle and poultry some 1,000 of these giving a return of 150 cals. Thus 2,150 cals are available in modern usage (good or bad) as well as lentils, root crops, nuts, fish, and grass fed meat etc. etc. Put in an alternative form 1 hectare of farmland (Indep. 9/2/01) produces 20 kgs of meat protein or some 300 kgs of wheat protein and a football pitch will feed 2 on meat, 10 on maize, 24 on wheat and 61 on Soya.

No one can fault the desire to reduce waste especially that between 7,000 and 3,000. Now these are figures of the 1970's and by 1980, the cereal production figures were greater by 30% in rice, 50% in maize and 40% in wheat ("A World in Crisis?"). Apart from the coal miner and navvy using the pick-axe and shovel who may need some 3,500 calories, the average male requires no more than 2,500, women and children somewhat less.

Thus on the surface there appears no problem in satisfying the basic need of all mankind for food. The reason why a billion or more cannot have more than one paltry meal per day, or that 800,000,000 children worldwide are malnourished (U.N.1996) does not arise from this cause. And the stunting of body and mind that results means that they contribute less to their society when they mature.

Man Remakes The Earth

The same conclusion was detailed in "Chemistry in Britain" as long ago as 1986, when examining the possibilities ahead and found that on 1986 bases were we to assume a steady population, of 8 billion, there would be needed four times the amount then produced, for adequate feeding. We must allow that cattle are to be considered as only 15% efficient in this regard even though they often use poorer soils. At current rates of yield 2.8 billion acres are in use, but with marginal cost an additional 5 billion could be usefully used to grow crops. Of this area of 7.8 billion, half could allow of double cropping i.e. we reach a level equivalent to some 12 billion, very much more than enough for world purposes. Were there double the present intake per person, there would still be enough, and the major uncertainty is whether in fact there is ever likely to be a doubling of population.

The world population of 1950 has doubled in 1998, but food production trebled, and we have far from reached the limits of arable land or man's ingenuity.

It would be good to reduce the level of waste since the energy expended in food production is enormous. There is the production of fertiliser, water to site, and the toll of dealing with parasites, fungi, insects and disease generally. These should not be allowed to flourish at our expense. One Third the energy expended on cereal arises in this manner, one fifth on fertilisers. Again fertilisers ,insecticides and herbicides can leach out of the soil, and may affect man. £120 million p.a. is spent according to the Times 31/7/04 and any reduction can only help man especially, where such leaching encourages the growth of deleterious algae. It is considered that half the nitrogen fertiliser used is unnecessary and in rice production 30% is wasted. However the newer techniques allowing satellite pictures of agricultural fields and sampling of small areas of soil can offer the farmer greater efficiency, lower costs and allow reductions in pollution of water supplies.

Since half of mankind lives by subsistence farming, improvement by use of the above can only enlarge the energy requirement and increases the possibility of damage to man and general pollution. EXCESSIVE use of fertilisers can increase the danger of salination on arable land, reducing its ability for crop growing. As the water table rises from man's demands the result is again increased salination. In fact during the last 20 years nitrates in our feed water has increased fourfold and phosphates eightfold. We know how to cope, if we have the desire.

There are said to be 6 billion humans in the world at the millennium and there are estimates that this might well grow to 10.5 billion by 2150 A.D. This possibility may need further attention to meet possible future demands. However where woman are freed to work and gain some independence in the domestic arena , the birth rate diminishes. In Kenya maternal mortality is 1% of live births, in the U.S. it is .002% and 137 million women cannot get contraception devices.

Malthus in the 1790's became concerned that the arable land of the world would soon prove insufficient to feed the hungry and could by over exploitation diminish returns on existing land. He wrote that the poor, often considered profligate and prolific and everdemanding must have there appetites in these directions curbed. He was thus able to support the view that charitable or local government help was a threat to society and where money was involved you had to be cruel to be kind if society was to be viable. Otherwise the growth in their numbers would lead inexorably to the worst travails of existing society. His message spelt out widespread disease, riot, and most importantly, since the French revolution had just taken place, a bloody revolution against the good, abstemious hard working upper classes.

His calculation was based on existing figures of life expectancy, infant mortality, and four viable children for each couple. On this basis the population doubled every 25 years. Nobody mentions today, when approving his rationale, that his forecast would become not the 37 million present in England by 1900, but 168 million and on the extension of this thinking the presence of 1.3 billion in the U.K. by 2000 A.D. A startling consequence indeed! Few mention that improvement of the standard of living could be an alternative direction to having too many poor children, a factor he himself did grudgingly accept as possible some 30 years after his first sounding of alarums. That improvement has led the Danes, the French, the Japanese the U.K., Italy and even the Chinese to reduce their rates by 1960 to below that necessary to maintain the population at the existing levels. Malthus did forecast for the longer term, but the Economist of 2/6/01 indicates that even a 1960 Netherlands surmise was 20 million for 2001 not the 16 million as counted. In 1969 Ehrlich, a biologist, forecast that in the 1970,s hundreds of millions would starve to death and not by war. These were the years of the "green revolution". Then by use of new hybrid seeds (not G.M. designed), wider use of fertilisers, pesticides and weed killers, famine ridden India became an exporter and China saw crop production grow by 2/3rds between 1970 and 1995.

Of course globally there are some 800 million malnourished but that is because they cannot afford to buy ,not that it cannot be made available!

To maintain the status quo in population numbers the offspring worldwide should number 2.1. The Observer 8/8/98 tells that in Japan it is now 1.4, the

U.K. 1.7, Spain 1.5, and in Italy, China, Cuba, Thailand it is well below 2.1. For India it is what was the level for the U.S. in the 1950's and the new projected maximum is now some 8 billion human beings ,certainly no threat in today's technology let alone genetic food developments. Bangladeshi women have halved their rate in one generation and the world average is now 2.7 and the trend continues down. (Times 8/10/03). The U.N. population fund tells us that in 1994 50% of the world population used contraception and ten years later it is 61%. So there are many reasons to say that Malthus was not correct in his forecast. He had been unable to set the future, just as computer models cannot easily weight factors today.

In all this, 24% of the earths land is fit for arable exploitation and 20% without irrigation. We actually have in use no more than 11%. Yet there are loud cries as a consequence of the woes arising from an ageing population. Offspring are the only security against want in the Third World. The greater the security, the lower the need for numerous children as a defence against starvation in old age or illness. In Africa and Asia it also needs knowledge and a loosening of the control of women by men.

Meanwhile 2/3rds of the Third World are malnourished and need another 6% (1998) to reach a minimum subsistence level. India is now self-sufficient in grain, yet while millions starve, soldiers patrol government sites housing 16 million tons of surplus grain. Bangladesh overall can itself supply 2,600 cals per person per day yet half the population has to survive on less than the barest minimum of 1,500 and that land could if encouraged, produce two, not one, harvests per year. China not so long ago had a famine situation almost every year for a population of 500 million. Today with a population of 1.1 billion there is no famine. Yet it is still a land of poor communications and where the laws of Capitalism have always dominated social life whatever the States name-calling. 80% of children in Mexico are malnourished and the cereal eaten by the rural population is less than that fed to their cattle for export. In Haiti the rich land has been allocated for single cash crops such as sugar, coffee to meet the demands for foreign cash for investment and meet the cost of repayment of loans from external banks and governments. The poor are allowed to erode the marginal land left to best meet their own needs. Columbia exports flowers and cattle officially and maintains its economy by drug production. In Senegal the monocropping of the peanut has devastated the soil. But an end to an endless catalogue.

Meanwhile (Times 17/3/05) there is reported that some £200 million in 2004 meant to alleviate urgent need were "sidelined", finding their way far from intended victims of catastrophe into the hands of corrupt officials and businessmen. In 2004 Iraq the $9.4 billion promised is also subject to corruption on a massive scale. Throughout the world but especially the deprived areas, corruption and bribery are the only tools enabling the use of necessities to get through to where it is needed.

Surprisingly Munich Re consider the cost of natural disasters below that suggested by the emotional traumas. Their estimate is $65 billion per year over the last decade, (The Times 12/9/05) where the G.D.P. of the U.K. is some £1,300 billion and the corresponding figures for the world vary between annual increases of 3% and 5%.

In Africa, the U.N. figures for agriculture confirm that with the possible exception of Mauretania, all are self-sufficient apart from the effects of war. But that is a very important factor. Indeed a letter from the Swaziland Solidarity campaign states that if the subsistence farmer growing maize had the same capital input in irrigation facilities as the commercial farmer, their yields would be 3 times higher. In sugar where they are most competitive they are beaten by the heavy subsidies of the E.U. and U.S. farmers.

The sea covers 70% of the earth's surface but for food only the continental shelving offers substance, below, the sun cannot penetrate and there is little life. Over-fishing has reduced the biomass of spawning cod off Newfoundland and Labrador from 1.6 million tons in 1962 to 22,000 in 1992. The only compensation is that today's cod matures twice as fast as heretofore. We still have not solved the problem where a quarter of the fish is too small, unwanted or contravenes regulations for the catch and is thrown back dead. Thirty five million tons, almost half the fish caught, becomes fishmeal, a quarter of this amount is fed in shrimp farms and the remainder to water farming and general agriculture. Incidentally the marketing of prawn farming has grown by more than an order and is now 700,000 tons. In the process mangroves have been severely affected but slowly measures to control and perhaps reverse these ravages are coming into place. There are also considerable pollution problems but again there has now been developed bacteria to feed on the resultant excrement and pollution of this sea farming.

A golden age is still possible, we just need to learn how far we can safely push nature. The F.A.O. constantly cries in the wilderness that better world-view management would increase yields some 10% to 20% in a few years, but we still need to ram home the fact that fish has diminished ten fold in many areas since the spawning of the industrial fishing era.

The Woe Brigade of scientists claimed that in the 1970's the Mediterranean was fully exploited, yet since then, the catch has doubled, perhaps due in part to the criticised increase in level of nutrient entering that sea from the land in the form of nitrates, phosphates etc.

Against, is the unfolding disaster in the Black Sea where such appears to have increased the mass of surface algae, reducing the penetration of sunlight. Worse still the decaying algae absorbs oxygen thus affecting the viable life of fish.

49 million tons of fish was the world sea harvest of 1965 in 1997 it became 110 million, albeit less of cod and haddock more of anchovy, pilchard and farmed products.

Where there is complaint that it needs several kilos feed to achieve one kilo of salmon and concentration of harmful residues can arise, this is the penalty for being at the top of the food chain and wild salmon too will concentrate deleterious substances. Where there is complaint that fishmeal leads to denudation we are pleased to note that anchovies, and sardine stocks have not deteriorated, but it is always possible to replace some of this feed from agricultural sources and there is enormous waste of nutritious fish not meeting the requirement of law or commerce and there are developing krill farms. (Economist 9/8/03)

Despite man's wastefulness and his mistakes there is no technical or logistic problem disbarring the vast majority of humanity from obtaining adequate sustenance. There is no requirement for charity. Charity does not grow food, it buys food already in existence and distributes it inefficiently. Famine relief operations, bank and government loans merely enlarge the problem. "World in Crisis?" tells that in 1961 Africa was 98% self-sufficient in food, in 1971 89%, and by 1978 78%., all done by the unthinking "help" of these agencies.

"The Club of Rome" in "Limits to Growth" of 1972 paints a similar picture to Malthus and how we love these messages warning us that the end is nigh. Its findings are however very sensitive to the complexion of the parameters chosen, the weighting given them and their interaction. All that has been given are as visualized largely by economists a first approximation to reality. They may not even include all the relevant playing cards and have sought no alternative to playing them within the present state society.

A Mr. Carr avers in the" Independent" of 12/8/02, we would be out of resources within 100 years, and oil would be out by 1992, now amended to 2030, but he reminds us, Jevons struck terror by forecasting the demise of Britain in 1865 by running out of coal. Today all minerals are cheaper and more available than they were 100 years ago, 50 years ago, 20 years ago etc. etc. and oil is considered as far more available than was forecast in the recent past, all because there is the ingenuity of man to get more of what exists than ever before. But how we love our Doomsters.

We inhabit a world that could be devastated by a large meteor, by volcanic eruption or according to "Equinox" 22/9/98 even worse, a hell's kitchen of flowing unstoppable basalt pouring from the innermost regions of earth. It has happened before. 250 million years ago 95% of life was wiped out, 65 million years ago 2/3rds of life and in the interim there have been the volcanoes. Their contributions dwarf that of puny industrialised man. India, Siberia, Scotland Greenland and the Eastern Seaboard of the U.S. bear grand witness to these past events. Their toxic gases have given years without sun over wide areas and reduced temperatures. In 1783/4 Iceland erupted and gave rise to 3 years of crop failure over much of Europe and the U.S., all by the release of dust and sulphur dioxide. In 1991 came the eruption in Hawaii of Krakatoa whose spew affected the atmosphere of the whole world.

Massaged into much editorial comment are the dangers of fertilisers, herbicides, insecticides and pesticides as in themselves evils, without regard to their use in unnecessary excess. We can cope with low doses of aspirin ,50 may kill us. Salt is necessary to life but 20 teaspoons can kill. Again one pint of water can save a life, 6 pints can kill. Any experiment on animals usually involves a very considerable overdose. The contents of every home and their maintenance in an adequate clean mode involves the use of suspect material and sometimes we do not know the level of dangerous concentrations for safe use and here concern but not ill considered rejection is justified.

Most pesticides will cause cancer in rats. Coffee contains more than 1,000 different chemicals of which 27 are carcinogenic to rodents. They are ORGANIC chemicals produced by the plant. 2/3rds of synthetic compounds are considered carcinogenic to rodents as are 2/3rds of NATURALLY OCCURRING compounds. The National Institute of Environmental Health at Berkeley University avers that one cup of coffee contains the equivalent weight of synthetic pesticide residues present in a years eating of fruit and vegetables by an individual. Of course this statement does not deal with the degree of deleterious effect of different individual compounds , but it should give light in some dark corners. For far to long has the myth that Nature is benevolent to man's needs ruled our thinking.

By and large synthetic compounds are targeted because they derive from the modern technology and the science of man as well as their use for food. But not all pesticides go into food. In 1977, one third of the usage of pesticides in the

U.S. was on golf courses, lawns and parks, half the insecticides were used on non-food crops. If all pesticide use ceased the loss due to weeds, mammals, birds, pathogens would increase the crop loss from 33.6% to 40.7%. In the light of present food productivity such a loss by itself could be taken in its stride except the few cases where the sensitivity to depredations were more severe. But it could make an important impact on profit! We need also to bear in mind that half the world food crop is produced with no such additions, largely by subsistence farmers.

Over-cultivation, whatever that means, can lead to soil erosion, too many elephants can denude and destroy a forest. The goat and consequent erosion may have destroyed the granaries for ancient Rome in N.Africa. Added to such considerations eroded soil sweeps into dams diminishing their efficiency.

Now from genetically modified programs has arisen the development of seeds which do not need the plough to effect adequate germination, an aid to erosion yes but not to the development of a program for the Green movement.

There has been established over many years the advantages to be gained from crop rotation and mixed cropping. Today such tenets are largely neglected because monocropping allows the more intensive use of Capital.

There can be brought to bear the introduction of natural predators, although there is the need to ensure they do not bite off more than one wishes them to chew.

We also have the possibility of the use of sterile male insects to reduce populations adverse to our being, and the use of insects and bacilli to act as predators on predators attacking sources of value to us. In 1888 a beetle species was introduced to California to control the scale insect in its action against citrus trees, In 1967 scarab beetles were introduced to Australia to reduce the dung problem because the home grown beetle would only act on marsupial excreta. A latest finding at Oxford according to the Indep. of 28/4/96 is a mechanism to release a foam to encourage swarming by locusts when there is no food. The saving potential is 100,000 tons per day and the cost of controls $350 million per year. There is a choice away from chemical pesticides, but not necessarily free of disadvantages.

Bracken, covering so much untended ground has now been found to have cancer potential for humans and the spread of rape over wide areas in recent times may contribute to asthma so there is quite a bit of nature to improve if we are to live fuller lives. Pollutants derived from man's activity and especially criticised when arising from his necessary industrial activity has been saddled by the "Greens" as the source incarnate of disease. Yet there has now appeared an intimation that not only in the field of cancer, leukaemia etc but even in asthma that attribution may well be incorrect. How else to explain a preliminary report from the Royal College of General Practitioners (Independent. 2/11/98) that although "pollution" has not been said to increase they find asthma has not increased since 1991 and since 1993 acute attacks have diminished by 25%. The little development of drug control, despised by so many, has hardly contributed. In the case of bracken there is a caterpillar quite able to control the spread of this unwelcome weed whose spread is otherwise so difficult to control and the caterpillar is under surveillance for unforeseen and unwelcome possibilities. Even the spread of the rhododendron into the many areas where it has become very invasive can now be checked.

In more mundane developments we see the development of bacteria to convert heavy oil into useful fatty acids, the use of algae to remove heavy metal as pollutants and for the process of their concentration for production. There have even been great strides in the economic production of desalinated water. Wherever we look we see the existence or near development of plenty, beyond plenty. For us there is a world to win if we adequately harness what is less than a stone's throw away. In this plethora of unharnessed knowledge the control of man by man alone, keeps us in the anti-scientific age !

In the U.S.A. a solution now put forward is to allow a fallow period of ten years for areas, previously in cultivation and now denuded, and adding sown soil-conserving flora. The solution is hoped to save most of 50 million acres and is economically viable. Appeals for humanity, appeals to plan against starvation are hardly relevant today if there is no pay-back on economic or political grounds. It is economic gain offering ever more profit that cranks the engines of motivation in this society. The only relevant plea centres that can be played in present circumstances is to diminish the excess of use by controls rather than blind regular programmes, needed or unnecessary.

Erosion, pollution, loss of fertility, arise most often in the modern world from short term investment, and correction of these phenomena can take a very long time and therefore tends to be left to government who are always strapped for the cash necessary even for their favoured avenues. Again the wide spread of listeria, salmonella and the threat of B.S.E. is accepted by the populace stoically and with little positive comment. Yet in Scandinavia salmonella has been virtually eliminated, as has scrapie in New Zealand and their farmers still make a living.

Certainly the British Government is much more concerned with producers and intermediaries in the food industry than the citizen. The latest example appears to be contamination of 5 tons of chilli powder with Sudan1 and the cafufling of the Food Standards Agency when detected by a French Agency taking 18 months to reveal the fact to the public thereafter. It must be said however, that in the likely concentrations released, this additive by itself is unlikely to be lethal. U.S. regulators banned the substance in 1918, yes 1918, and as a result of further work it was declared a carcinogen by W.H.O. in the 1980's. Its allocation towards the offer of a nutritious feed for our young of 37 pence a meal on school food intake shows the real strength of state concern. (Sunday Times 27/2/05)

Preservationists tell us that 90% of the world's food relies on some 20 species, and that in these circumstances disease or other problem could cause catastrophe. In such circumstance the understanding of the means of altering or controlling the gene structure quickly, needs to be paramount and not subject to Luddite treatment.

While the present development and possible neglect of unwelcome attributes are there for all to see, the cause is ever the temptation to maximise profit by too speedy a launch.

In all these circumstances less than 2% of research and development is spent in the areas of agriculture and the environment.

Today in the U.S. and elsewhere the use of antibiotics in cattle production has become general with legitimate forebodings as to the dangers to humans of resistant strains of germs arising from this intake. But the cry of the E.U. against the use of this technique for milk is not based on such a consideration. Within their territory, they are almost drowning in milk and they do not want any more or the competition it might bring in its train. Its use could only upset their price arrangements.

Fashion and perhaps health, too plays its part in that antibiotics can reduce the fat ratio in meat.

Farmers have always manipulated the genome but because nobody quite understood that they were manipulating gene structure it was acceptable. Now that we are beginning to understand the nature of the endeavour all manner of obstacles are put in its path by religion , "guardians of ethical codes", and malcontents. The slower breeding techniques of the past has given cows with high milk yields, cereals unrecognisable, in yield and acceptance over their grass forebears. With antibiotics we have available, lean pigs, even the ability to change the ratio of saturated and unsaturated fats making up the constitution of the animals on which we feed. Present-day husbandry has been a promising storehouse for a cornucopia, but these changes, have largely blinded us to the loss of good important, previously innate characteristics to which there is no return. Even the Green revolution producing crops with smaller stalks and heavier fruit may have lost something , but it did solve the food logistics problem for much of the Far East. However the tardy pace previously necessary has given time to assess and eliminate unwelcome products. Speed in development could also increase the presence of infective agents and their carriers hitherto restricted in range. Just because it takes 12 years on average to cross-breed an animal successfully and say a year to produce a G.M. product, it should not become mandatory to waste 50 years to ensure safety and this is the direction we are taking.

Remember too, that these earlier techniques included the use radiation and chemicals. More than 2,000 types of crops were treated with gamma rays to produce mutants some of which are now offered by organic farmers. (Economist 13/3/04).

In this treatment may we insert the already mentioned effect, that were a global warming to take place, the the earth would become more fruitful, but without man's attention it might just favour the weed as its more natural product. It might also encourage insect life e.g. the malaria ridden mosquito, which has been a stranger to our shores since our wetlands were drained.

The hopes are waning for the promised glorious flow of cures from deadly diseases by forest magic ("Economist" of 20/11/99). Of 35,000 samples from 12,000 plants only 3 significant products have been discovered. The Amazon is not teeming with such cures. Some there are and after all half our cancer drugs and about 100 others developed since 1960 are plant derived, but costs and site difficulties in the Amazon and other such sources must be high. From a Chinese herbal medicine has recently been synthesised a replacement of DDT and thus man using his ingenuity on a natural material can be a great benefit since the use of DDT has been banished. The Times of 10/6/05 mentions a fungus which can when sprayed largely eliminate the mosquito. the spores kill the insect before it can spread the disease.

In our interest in Witch Doctor cures one can forget that his area of practice is practical and is largely concerned with his local parasitic infestation problems not with cures required by the civilised world. They would tend not to be concerned with heart disease or cancer. The possibilities opened up by genetic engineering offer a speedier advance. As one example the development of the hairy potato which repels the aphid and the diseases it may carry, should not be repelled by mankind.

Going back, can never be going back to reality. Old valuable paintings are difficult to appreciate if the fine details and original colours are not restored. Modern hands and arts alter the original, but for most there is more beauty to be seen after such amendment. The Artist too may have preferred the amended rendering to the deterioration by time if his aim was to please society.

It is possible to think of life without the arts or even football but not without food.

Apart from the imperative necessity to ensure a best future by scientific assessment, of any additional and important contribution extractable from "Green" propaganda, waste is unforgivable with consequences for those yet to come. To this end the "Greens" contribute well.

Present detergents used are not the great danger projected by the Greens. While the phosphates used can if they find their way into rivers encourage the growth of algae to an extent that it interferes with aquatic plant, insect and fish life, their alternatives also contribute adversely.

Today, the main source of unwanted phosphate and nitrate arises from leaching of agricultural land where it is not uncommon for half the fertiliser distributed to be unnecessary and an overdose. It is this enormous amount that leaches through the soil by rain and in its very nature of distribution cannot like detergent residues be treated.

The remarkable advances of man in the supply of food has been accompanied by a considerable diminution in the varieties available and it is only recently that we have had availability of lesser known grains although at greater price. Wheat, American corn, soybeans, rice and perhaps oats are present in overwhelming quantity and at consequent lower prices to supply our needs. Because we understand better the vagaries of these crops and through familiarity we are more likely to get the best out of them, the polarising effect of concentrating on fewer and fewer grain sources is accentuated.

The result is a reduction of our gene bank. Were we to enter a phase where the result was an initiated inbreeding weaknesses then we would indeed have problems and such weaknesses may already be present in our edible plants and animals. Any disease affecting the major plants used could because of our widespread dependence on few sources create havoc. Insecticides, fungicides and pesticides may help but we need ever to ring changes to ward off resistance development by pests. G.M. food may help in this struggle by offering other options even were resistance to arise with them. Although the effect of these on the consumer is much exaggerated it might not always be negligible and there is the manufacturing end to consider where the larger concentrations could indeed cause adverse effects. Today, our society has only best yield/costs ratios in mind and only changes to open a new market or promises greater profit. In fuels the same concerns are targeted, but with the plethora of food now logistically available, other considerations must edge their way forward. Efficiency, yes, but not at all costs. Let us accept a less efficient source of fuel or agronomical technique where it is affordable. Perhaps now and only now, can man introduce or perhaps re-introduce the qualities for disease resistance. There is the need to be looking to improve their food values, in the best combination of protein, carbohydrate, vitamin and mineral assemblies. There is the need therefore to know more of their nature and the best proportions for intake. Capitalist mass production has been concerned with size and appearance and in optimising these has casually accepted the major elimination of taste in every sense.

The history of the selection and alteration of his food reflects the history of man, and for 10,000 years we have been improving on nature by means "not intended" by that goddess. Agriculture is about destroying the previous "natural" environment. The present countryside is quite unlike that of feudal times even of that of the Georges of England.

We have uprooted hedges, drained wetlands, denuded the forests and ploughed heath lands. We extract, we separate, we change the nature of food. We render the inedible nutritious. We are all chemists when we cook, and more than half the population engages in this activity every day with little fear or disfavour. We have no wish to return to the days before the discovery of fire, whoever recommends it.

Mendel in his researches offered the conclusion that crossbreeding could be explained by the presence of discrete inherited parcels and when Crick and Watson deciphered the genetic helix structure of D.N.A. in 1953 the cupboard for the background knowledge was made much easier to see. The new technology based on gene structure allowed a much faster rate of changing our raw food and for this reason there needs to be great vigilance if no important damage is to be fixed permanently. The stage has been reached where with some difficulty we can introduce or subtract one gene covering a few characteristics and no longer need to suffer an accompanying loss of characteristics worthy of retention. If we are successful in transferring the gene allowing legumes to utilise nitrogen in combination with bacteria in the soil, then we can avoid or much reduce the need for nitrogen fertilisers. If a gene introduction allows reduction of compounds otherwise necessary for protection against depredation there is no reason why such should not be advantageous and used.

The crossing of genes from similar species should spell CARE rather than great danger. In Canada 73% of rape production is Gene Modified and already ensures that related species especially of weeds are kept well apart.

Along comes an article (Sunday Times 12/8/01) telling us of the danger of creating superweeds, not super plants. Apparently if local farmers each purchase a different gene modified rape seed and they all managed to breed all their genes within each of them their could be transfer to a weed ,producing a super rape weed. The author obviously does not realise how difficult that total transfer to all would be! And were it even possible there are already patents for producing seeds that do not reproduce, just as with flowers.

True there has already been conferred herbicide resistance to a wild radish, and some weeds related to oilseed rape but one can change the herbicide. Against this feature we have a pig farm (Observer 25/7/99) with a few hundred animals giving huge quantities of manure and ammonia. the deposits of ammonium sulphate stripped conifers and affected the other woodland about. Why then should we deem the latter a better approach to our food production. Especially because it is a new field the sure touch of experience is not always present, but this is not a reason to fully revert to old time procedures. What is needed is CARE!

Commercial growing of G.M. crops since 1996 has as yet shown no danger to health at 2005 and in 2002, 34% of the U.S maize crop, 71% of cotton, and 75% of the Soya crop in that country were in that category. Canada, Australia, China , India and Indonesia are all proponents, especially in cotton. In Britain there is resistance even where Ms Becket has been pointed out that maize has no wild relative here that could possibly take up such genes and that G.M. maize in the trials were found to be kinder and safer for the environment than the non G.M. product today, (Economist 15/1/05) Figures arise of 80 million hectares planted with such seed, including some 30 million in developing countries. If there were a wolf among the flock he is now roaming freely, so far without showing wolflike tendencies. Considerable savings have been noted in the economics of growth and even more in the use of herbicides and pesticides. (Econ. 26/7/03)

Between 1999 and 2001 the earnings for China per hectare when using G.M. cotton has risen by $500 per hectare and although the patented seed is more expensive, it allows an 80% reduction in pesticide use. G.M. techniques have already offer us virus resistant sweet potatoes in Kenya, drought tolerant barley in Egypt and protein enriched potatoes in India.

The recently published work on potato lectins by Dr. Putsai far from proves his fears. In a small sample of rats, he noted some thinning in the gut by ingestion of raw potatoes which should in any case be eaten by man only after cooking. Replication was poor and remember the feed was 100 times the concentration present in normal G.M. tubers. The introduction of a freeze resistant gene from fish to plants poses a more unusual combination and needs more careful watching, but not in fear and trembling.

Unfortunately the first commercial developments offering a herbicide or insecticide packaged with a grain capable of resisting that compound has dominated present work. It is not itself bad but can be seen in the larger sense as an aberration. We can now introduce improved foods in breadth and depth, foods more capable of resisting adverse conditions in nature. It is now foreseeable as possible to activate, inactivate or delete genes and more questionably, to transfer them across species. However the transfer from a fish of cold resistance by G.M. techniques ,to tomatoes and strawberries has as yet shown no harmful results on humans.

It is only since this field of knowledge has been pursued in scientific fashion that the hounds have been baying loudly. Only now that Science has stepped in has the process become a source of fear and not standard practice. Genetically modified strawberry uses the gene from flounder to enhance frost resistance. There is nothing wrong with extending the drought resistance, decreasing the rate of deterioration prior to or after harvesting, increasing the ability to cope with salinity, or introducing disease resistance by gene implantation.

There needs to be the proviso however that it is adequately monitored against other deleterious effects, and the product retains a balanced appeal. We have offered G.M. rennet to cheese lovers who are vegetarians for many years and yes we do realise that the rennet does not enter the food chain.

We should also look with favour on the use of renewable starch converted by G. M. yeasts into polymer products previously only available from oil and there are others. To the extent to which we are concerned with the role of carbon dioxide, an agricultural source that absorbs it rather than expels it is should not be so lightly rejected.

We eat strawberries as well as flounder. True the new combination may have unforeseen consequences, but then let us test and test again. There is no longer need to suffer the tedious pace hitherto required for this to happen using methods, venerated only because they are old and which can only shuffle the gene mix around often eliminating the useful with the gene desired for rejection. The Peace rose took some 20 years to develop with many virtues but no fragrance. Today the unfolding vista is to have a beautiful disease resistant rose, one that smells like the rose of yore, and the like with smell, taste and appearance of food as well as medicines without side effects. How elegant and worthy is the simple introduction into cotton of a blueing gene, when it saves the production of a dye and the business of dyeing for that colour. Why should Man not develop genetically modified birds or animals to produce say, insulin to avoid injections or produce human milk in cows, which must be an advance on cows milk. Indeed why not introduce vitamin A into rice and save millions in the East from blindness against those who would reject reason.

In order for the gene to penetrate a cell, the present technique requires a carrier which can anchor it to a chosen site in the DNA structure, it also means a variation however slight in the size and structure. The bacterium of choice at present is called Bt, commonly found in the soil and it is to be mentioned that it is in common use as an insecticide by the "organic" fraternity,

There is also involved chemical and physical changes in the mode of adherence of the gene to the remaining structure. Then there is the activator which is needed to bring the gene to life and there arises a possibility that the new configuration could lead to a "misfire" or a "firing" on a differently sited gene which might therefore lead to a differing response from an already established gene. There could also be a stability problem and a coiling problem as well as different modes of behaviour arising inside and outside the cell. If the "junk section" of the chromosome accentuates or attenuates the change made a problem might again arise. There is also the possible transfer of allergies say by using a certain nut gene. There is much promise in examination of such features and no cause for rejection.

There are many offerings and one needs to separate the chaff. If potatoes can produce oral vaccines, if sheep can produce in their milk, the human clotting factor against haemophilia, we should be glad of it. Where there is the promise of fibres, rugs, chemicals, building materials from renewable resources we should be happy to accept such promise!

In all this we should remember that while rats and human beings are mammals, rats are not human and will react differently in quite a number of areas.

Man Remakes The Earth

A mating between a donkey and a horse gives an infertile mule and if genes were easily garnered by plants and animals then present diversity just could not exist. It is very, very difficult to breed within a genus and nearly impossible by natural means to breed outside it.

English Nature tells of 3 G.M. rape crops in Canada, grown almost contiguous to each other and made to resist 3 different herbicides only to find after 3 growing seasons some were resistant to all three. So one should not grow crops with different determinants together is the correct answer ,but not that given by them. In any case the variety of herbicides and pesticides available to control this situation are great and all should recognise that life in nature is struggle against nature or with it. In any case, our foods would have short shrift when untended in nature, e.g. rice and rape would disappear in five generations without the use of herbicides, and in all probability their descendants too.The danger posed of monster weeds etc are very unlikely. Even so we can cater for that contingency by not growing weedy relatives in the vicinity, a practice already extant. As a scare ,One monster weed , a hybrid from oilseed rape and sinapis arvensis , was put forward as a poison in 2005 by the press, a horror of horrors . Yet closer examination revealed that far from displacing food sources, it does not even produce viable seed.

Genetically modified American wheat as grown saves 3/4 of a billion dollars yearly and has been eaten widely for certainly 5 years. The claims of increased yields in the countries grown are, China 23%, Brazil 24%, Philippines 41%, and in the U.S., Argentina, S. Africa the savings indicated are 5-10%. Today more than half the food consumed in the U.S. contains G.M. ingredients.

The bee and the ladybird were not seen to be affected but the monarch butterfly was said to have suffered grievously. "Scientific American" has looked into this question more recently in its issue of 14/9/01, and found that Monarch caterpillars were affected only when 1000 grains of the pollen covered a square centimetre of caterpillar food. Only one commercial modification was involved and was withdrawn. There was no toxicity found at the levels arising in nature, and for good measure the still higher levels used in experimentation seemed only to affect one strain. But it is necessary to judge whether such changes in protection are better or worse than that which exists since Monarch species is affected by the insecticides used otherwise.

There is the knowledge that botanists have underestimated the number of plants by as much as 40% which would add 100,000 species to those known. (Indep. 1/7/02). Previous insecticides in use also affected bees and ladybirds. The finding seems to be that the function of the digestive tract of the butterfly is affected, i.e. nothing is triggered off in the body proper. We would remind the reader that we have not the same digestive system and are thus likely to be differently affected. Again in Arkansas, farmers till 1 million acres of G.M. cotton previously decimated by the bollworm. It also resists aphids so that although it does not affect ladybirds, it does diminish their food supply. However they use little of the organophosphate pesticides previously necessary. In this we can see that with development, crops can be developed to choose which insect or even bird shall be preserved if the will is in place.

There is also the important claim by Monsanto that its "Round up" herbicide binds the soil and by reducing the need for ploughing diminishes erosion.

"The Independent" of 20/5/99 discusses the deaths of the Monarch butterfly where ingestion of G.M. maize pollen has been posed as a major cause of their decline and threatens their demise. Dr. Munro however suggests that the cause is largely due to deforestation in Mexico where they overwinter. This concern has lead to a further appraisal (Econ.22/9/01).The National Academy of Sciences of the U.S. produced 6 papers by 29 contributors supported indeed by an industrial body but which has concluded that it can affect the larvae of only butterflies and moths. If we know this much then the damage to insect life is restricted and it may not be impossible to derive a product which is more specific. In many cases the pollen tends to collect in the middle leaves which are not a site for feeding.. In any case nobody is suggesting as an alternative, to protect species of insects, that crops be grown unprotected against them and we know that standard insecticides have a wider range of effectiveness over many of them. (Econ. 20/10/01). In any case (Indep. 14/2/02) tells of research showing that the death toll by freezing during its migration cycle as highlighted previously, is twice as high as the total population. Which means that all figures related to insect and plant numbers and species need better establishment before figures are presented as ultimate truths.

Improve yields and there could be less need for deforestation or environmental damage. Again, it has been found that most wood pigeons avoid pesticides on crops (Econ. 4/9/99) until the viability of the herbicide is much reduced and the lower the amount used the better all-round for health, and that is the promise of gene technology.

Were the procedures not adequately controlled it could happen that a gene which creates danger to allergy sufferers, say from nuts or beans, could affect a wide range of foods by the gene's introduction into other foods but that would certainly be the height of irresponsibility. There is no reason why we could not also introduce a gene to block the causative protein, and enable the wider enjoyment of foods hitherto a source of suffering to so many. In Japan a majority suffer allergenic problems with rice and there is now the capability of blocking the production of the allergen. Where as a result we reduce the use of herbicides and pesticides wild life may increase, but perhaps differently.

There is in the offing the growing of renewable fuels, even the making of plants equivalent to meat in all nutritive respects. There are even vaster medical rewards, but the Luddites are out in force.

All this means that the problems are not doomsday problems but need close care to assess and where necessary remedy the risk. Any point of risk should be left on the research and development shelf until a decision as to its resolution has been taken. Many of these will have been answered, but due to private ownership remain outside the public domain. It is control for the purpose of profit and nothing else which makes this wonderful development a danger. Despite this feature modified soya has now had some 10 years of trials plus 5 years of enormous commercial production without health problems.

The reader is again reminded that in all this discussion the problem of an adequate supply of food for the world population has already been resolved, and there is no bar to the use of biological controls other than ensuring that the introduction does not enlarge the problem area. In California the leaf hopper has developed into a more voracious species on the grapevine. The Mexican wasp has been introduced because its larvae is laid where the leafhopper deposits its eggs and the larvae destroys them.

The modern development is only to be encouraged and it is theoretically to the good that feeding the hungry of the world is a social problem not one of technology, since its solution becomes more obvious. Profit is therefore the only reason for the rush!

Already there is a patent situation preventing the use of seed from a previous harvest and there are terminator patents to prevent evasion and for rendering harvested seed infertile. The Monsanto's are not giving it away but ensuring that a yearly tribute is exacted. In these circumstances the wider spread of genes to other plants can now be controlled.

When you transfer a gene from an edible fish, plant or animal to a human food source then after digestion it should do no harm, but one must always be aware that a new combination may contribute more than its parts. Were the gene transfer to take place from a source capable of producing poisons then even greater care would be needed in its assessment. The matter of producing edible food is quite different from gene incorporation in our basic structure by targeting. Of course it is possible, but then many things are possible.

The new ball game has breathtaking promise for our future in its promise of extension into growth areas, previously inhospitable. As mature beings we should wish to learn the new game however exciting, when every test of method in this game shows superiority over those in play today. The way is open for us to vastly increase yields, storage time, taste and quality.

It is the rush to announce, to publish, to patent, and to neglect adequate evaluation which has largely contributed to disasters in the field of drugs. The fault lies with the social system which allows the question of control and release to be channelled to individual groups or persons allowing their claim to the usufruct and leaving not a crumb behind. It is part of the "first past the post" game of this society.

Thalidomide was introduced without adequate key tests on animals or people. Now its properties have been realised it is still a promising material for treating leprosy and some cancers.

We can now introduce new properties into our crops and animals and we have begun, because of the contribution of science, to explore with an intelligent eye all the varied products still in existence which have not seemed to merit greater attention up to this stage. It is science not "things my grandmother taught me", that has highlighted the need to maintain and stock the genetic pool of existing species. It is science that has enabled the replacement of materials previously only available from animals, but now to be found in plant origins and we are near to making our own gene bank and building a stock for the betterment of man. The fears do have justification in that our society is concerned to advantage certain small groups rather than society as a whole and indeed would accept disadvantaging the many if thereby there was wealth to be derived for the few. Today we have genes introduced to better resist the effects of pesticides. What we need is genes to eliminate the scale of need for fertiliser, insecticides, pesticides ,which cannot be transferred by any means to weeds and their like and contribute positively to our food bank.

Legumes are host to bacterial colonies which supply it with nitrogenous feed and it need not be beyond the ken of man to safely transpose a similar gene structure to other plants while maintaining their individuality.

However, there is in this the overriding requirement that the means of life are not controlled by the few, if preventable cataclysms are not to arise and man's future made safe.

Let us not be totally deflected by siren calls to preserve the countryside as it is to-day when its structure is only 150 years old and in that time the insects have had a field day proliferating as only insects can. That countryside arose out of enclosures which largely eliminated the small farmer. The Yeomen of England were no longer that useful to the powers that were. The "Greens" as usual, where they do not exercise a nostalgia for a fanciful past, seek at least a status quo position on the insect population.

They do however signally fail to make us aware that the present insect population, like the rat population, is the result of man's growing efficiency in nature and that most in their nature compete against us for the food we grow. It is the result of man's efforts that the insect and rodent population has been able to increase its numbers probably hundred fold. The Greens should tell us the levels of preservation desired, whether at today's level or that of 100 years ago. Perhaps they should let us have a view concerned with the preservation of lice, fleas, bed bugs, cockroaches or mosquitoes but their expression illustrates the wish to preserve insects and neglects by contrast the many humans in want. Perhaps, because he is not in that position.

One returns again to the social fact that the logistics for supplying food and shelter for mankind is already solved even for the foreseeable future and therefore field insects could continue without harassment. They should only be allowed to continue, within the rules concerned with our viability.

This additional technique in its presentation is an added safeguard against unforeseen calamities to our present supplies. It could solve such problems as to the most nutritious food, in its ratios of meat/ saturated fat/ unsaturated fat and reduce waste. It might allow us cope with a colder wetter climate by still producing wheat in less hospitable climates. We verge on a plethora and if that be our aim, there is still no need to rush into the unknown. It cannot be repeated that care needs to be taken, especially where profit is the main motivator.

One should also be aware that many root plants including the potato, and the bean as well as the nut would be forbidden if newly introduced today, because they can be poisonous or give dangerous allergic problems. Sometimes nothing ventured, nothing gained.

Man Remakes The Earth

The smell of sulphur arising from boiled cabbage is no longer with us, but this has not necessarily meant that the newly derived specimen is better or worse for our health. We are told that Syngenta have available Brussels Sprout seed that will taste sweeter. As a result of better G.M. techniques they are able to improve the taste problem and not much else. The pampered crops we eat would in no way stand competition from what we are pleased to call weeds, or their forebears. As with domestic animals they are the result of patient selection when this was not understood as gene selection and measurable only from external differences. Our produce needs man's constant attention and it is this factor which has caused the development of pesticides, insecticides, fertilisers and fungicides. It would indeed be desirable to introduce into the innate structure, the means of obviating their use. But we can at least reduce the amounts used significantly and allow the use of less toxic chemicals. It may also allow a saving from enormous costs, e.g. in the U.S. more than a billion pounds are spent on such compounds yearly.

If all plants could fix nitrogen as do the legumes then nitrates would no longer be a threat by pollution of our rivers and waterways. Advances have been made in engineering the banana to prevent hepatitis B, and the tomato as an anti-dote to rabies. Perhaps we could dispose of the refrigerator as a preserver. Perhaps there could be introduced by these means a high wheat protein crop containing all the values we want in meat. The means for transfer of genes is close at hand coupled with the real fears that where profits rule such crossings would not be properly monitored and might open a recipient to new diseases. B.S.E should be the warning not against innovation but careful testing. We have come a long way since the vehicles for such transfers were rudimentary and could themselves pose dangers.

There are already in existence plants, indigenous to arid areas with large roots brimful with high grade starches and some of these can survive very heavy frosts down to -20 C. There is a species of Vernona a common annual herbaceous weed that has been hitherto totally neglected and which tolerates arid conditions and high temperatures. There is the cactus gene structure which may also help. Its oils apparently have niches where they are superior to oils in present use. Some have high protein contents and speedy life cycles, some contain the more important and less widely distributed fatty acids. Their yield may need improving and it may be desirable in any case to introduce any good qualities inherent in these plants into our more staple foods.

"The Independent On Sunday" 10/6/01, tells of the macuna bean which leave good residues of nitrogenous fertiliser in their decay and have enabled the tripling of maize crops and enrichment of soil in S. America, and there is more to be discovered. A recent B.B.C. 11 "Horizon" issue of 19/12/02 has suggested that part burning waste to charcoal level, could make a considerable contribution to fertility of poor soils such as are available in the Amazon tropical forest belt. It suggests such as the cause of great fertility enabling the growth of civilisation in the Amazon some 2,000 years ago.

To date the soil is the bedrock of our life and therefore erosion and the causes of erosion needs to receive more urgent attention, and the simplistic explanation of scientists and do-gooders do not always give the full picture. It is sometimes necessary to dig deeper. As firewood becomes scarce, dung and other residues are used, which would otherwise be composted and used as fertiliser, and as already stated we are near the stage where seeds can be germinated without the plough. Without the plough there is a saving in its fuel costs, water and non disturbance of the soil equilibrium.

Plants do not imbibe dung or faeces. Certain insects do, and sometimes animals. Plants imbibe by osmosis and use those compounds from any source possessing a size to pass through their membrane structures. They include the salts they need and could include inimical substances if small enough and soluble enough. It is now largely recognised that the use of natural manure helps structure the soil to the requirements of plants and that is its contribution in addition to its content, but it is not necessarily the only or best solution. For cultivation there is little wrong with hydroponics except cost and ease of availability. Our present day knowledge may not be 100% in regard to nutrients, especially rare elements, allowing us to act in full safety. Natural fertilizers however can only draw from the soil on which they feed. The problem of artificial fertilisers is largely controlling the excesses and ensuring that they do not by leaching, enter sensitive areas of the environment. However, for those who have eyes to see, the problem is just as applicable to manure fertilisers, as has been shown in Holland. It is not as common because there is so much less about. Not even the Greens seem to envisage increasing the domestic animal population just as a source of fertiliser and their concerns seem to be with feeding themselves.

The growing alternative posited among the more prosperous natives of the Western World is "organic" farming involving the exclusion from the cultivation arena of man-derived chemicals in the form of fertiliser, and insecticides etc.. Organic foods are not tested as vigorously against fungi contaminants and carcinogens derived especially from mycotoxins such as aflotoxin. A century ago these abounded in the general food chain and modern intensive farming was only acceptable because when accompanied by effective control it was safe. We forget at our peril the now nearly or entirely eliminated diseases resulting from modern hygiene, modern farming controls and modern medicine. The ancient presentations were just not good enough to cope or there would have been no change. But memory is short lived and we are often seduced by nostalgic thoughts. The production of food is a prosaic enterprise. There may be monsters lurking in the shadows and we must take account of them, but there have always been monsters and any return to a previous mode of production deserves much more suspicion than has yet surfaced.

Prof Trewavas, a plant biochemist in Edinburgh describes the fears of those who look forward to the continued effort for control of man's food by man. There is a general perception (Sunday Times 12/8/01)that we produce good food but dangerously adulterated by nasty chemicals and worse. He notes this alarmist suggestion is unchallenged by an effusive, uncritical press. He stresses that conservation is not going back, but managing change to conserve the viability of man. In other words it is not about keeping insects, diseases, vegetation or animals, at their present man-made levels or previous levels but the conservation of man. Prince Charles lives on a different plane from others and if he wants to rear plants that love music he can indulge his tastes. If he does not like modern architecture he is free to say so. He is a man comfortable with the past and unlike those of us who form the vast majority, the past has served him well.

The wilderness may be important, but not we suggest as feeding the hungry. Yet organic foods can only increase the land necessary for cultivation and that must mean less land for forest and wilderness.

Against all these projected odds, we live longer, and we are beginning through science to recognise the mode of operation of food in disease contribution and health contribution. The more we eat of the right conventional food including fruit and vegetables the better will we withstand the ravages of time. Sir Richard Doll tells that only 25% of us eat the right foods. It may be that most cannot afford them. In any case we know that what is considered as well tasting food differs widely and blind testing does not confirm that organic food tastes better. It may just be that fresh food appeals more to our tastes, that food species grown for ease of gathering or increased shelf life do not enhance their appeal to the taste buds, but this has little to do with their method of cultivation.

The majority of "organic" food is imported and subject to little supervision as to mode of production. Prof. Krebs on examination has not found organic produce healthier and Lord May ,President of the royal Society agrees. the Advertising Standards Authority has also challenged this aspect. Certainly fresh food grown in an allotment or garden tastes better but only because it is fresher and we cannot be sure what is implied by the term better tasting especially as "blind "tasting does not bear out claims made.

There is developing a critique for supplying our food with the minimal requirements for good healthy yields but every day in Wisconsin livestock produce enough in manure to fill a 76,000 football stadium (Econ. 20/10/01)and how do we control that effluent. It contains too much phosphate, and its diffusion into lakes causes the growth of algae.

Diet intake is considered as a cause of 1/3rd of cancers ,so what we eat is very important. But bread contains the carcinogen furfural, potatoes 2 nerve toxins but all at concentrations to low to matter. Every day (see Indep. 30/7/99) we ingest a 1/4 teaspoonful of carcinogenic material, 99% of which arises from natural plants. Taking "non-organic food means ingesting from that source, 1/20,000th of a teaspoon of pesticide /day and the safety margin, and that can be argued, is set at 1/100. The naturally occurring insecticide Bacterium Bt, used in 1/3rd of U.S. crops have now brought on resistance and this material is also widely used by "organic" farmers here. However it is likely that different strains can target different insects or the same insect and what is gained is an increase in the armoury of man against privation. Organic farmers use modern developed seeds not those found in nature or derived from their previous cropping since otherwise disease and yield problems would be aggravated. In any case since the yield is up to 50% less than by other methods further development would need more farming area with its inevitable repercussions on ecological pressures. The modern oat, wheat, barley, tomato, turnip, sugar beet or potato etc. has been genetically manipulated away from past produce of even 50 years ago. "The Economist" of 22/2/03 tells us that herbicide-resistant G.M. sugar beet on trial in Britain would save some £23 million per annum and allow a lesser use of agricultural chemicals and "The Independent" of 13/3/03 that the consequent reduction in weedkiller has allowed a tenfold increase in weeds and two fold increase in insects with no deleterious effect on yield. Thus, we can play games altering ratios. We can choose small weeds against large weeds to please those who are concerned to preserve weeds.

The use of cow and pig manures has spread the likelihood of dangerous infection by E.Coli 0157 and other dangers that lurk in pig intestines and these could only be avoided by adequate composting or heat treatment of such manures. According to the Times 9/4/04, organic ham contains sodium nitrite. In some countries this is considered as carcinogenic and may also affect hyperactivity in children. The use of copper sulphate in organic farming has had to be stopped because of its effect on the liver. For them an organic egg arises from conventional commercial birds fed for 6 weeks where 20% of that feed is not organic, their insect control remedy, Bt can result in lung problems, and their use of rotenone has been implicated in Parkinson's disease.

Already victims have been claimed (Daily Mail 15/5/2000). In one year 100 cases appeared in the U.S., with deaths arising from eating lettuce and strawberries, 2 in the U.K., with one death from goats cheese. The "organic" approach does not seem to protect from pesticides as according to " The Independent" of 1/12/2000, pesticide were present in "organic" baby foods above government limits. More recent work (Indep. 20/11/02) mentions that a study for the Food Standards Agency finds 56% of factory reared chickens are infected with campylobacter, but over 90% of Organic and Free Range birds are so affected. A Previous Danish Study came to the same conclusion. Fortunately we are beginning to use bacteriophages and for example one is available which will kill 16 out of 18 toxic strains although it will also destroy 8 of the 73 in this group which are harmless. (Independent 24/4/03)

Apart from the pesticide problem it was found in 1995 and 1993 that 13 children were affected in Germany by eating "organic" foods, two dying. And we must accept that these are early days in the use of "organic foods "There is no doubt that pesticides do screen out other dangers. Nature feeds us but we need to sift the wheat from the chaff and that applies to all farming since there is no purpose in nature concerned to attend to our food needs.

Far better to concentrate on alleviating the misery of 2 billion anaemic beings causing many maternal deaths, far better to introduce means to increase the iron content of rice. Far better again to increase its vitamin A content whose consequences afflict 250 million children with blindness. But do it carefully!

In any case the G.M.horse has already bolted, soya as an example has been widely disseminated in foods for more than 6 years. There is little amiss with our present knowledge in transferring the gene from rice offering a few plump grains to advantage another rice with many grains which are however small. What have the other more laboured breeding techniques to offer which is superior, except sometimes important room to observe at a more leisurely pace at the cost of losing as well as gaining in qualities? Much of the rest is politicians dumbing down to their electorate and the green Luddites.

In any one year, sands from the Gobi reach the U.S. and those from Sahara reach England. This is one world which just cannot be parcelled into neat compartments for the convenience of those who support "organic" farming. According to the Times of 9/4/04 organic farmers are allowed to use 17 compound fertilisers, 7 trace elements all made in factories. They can also use 2 soil inoculants, 6 sulphur and 8 copper based fungicides as well as 8 insecticides. One insecticide is harmful to fish and Bordeaux is considered as toxic, but of ancient origin and therefore respectable.

We have brought to attention our criticism of solutions derived from general possibilities without attention to the particular features which may or may not be of importance. Some new farming projects in Africa have been a disaster, similar projects in India a success. What can be seen with hindsight is that much depends on the social organisation, how far the relevant infrastructure can cope and the political structures and intentions. Asian sites have benefited from irrigation, fertilisers, high yield seeds and multiple cropping. Of major importance was it that in India, land is limited and labour very available and very cheap, whereas Africa offers cheap land and a scarcity of labour. In the areas of concern the latter live largely by wood clearance, moving on once fertility declines. The problem for Africa as well as parts of S. America is the pressure for cash crops, largely absent in India and China. Africa also suffers more of drought, disease and salination and cash crops acerbate the problem.

In these lands of poor agriculture and of mineral exploitation, farmers are paid little for their produce and therefore tend to produce little above subsistence level. They have plenty of free time and efficiency is of no major concern. They are bedevilled by "free" imports, charitable in origin from lands with much surplus. The practice ensures that home grown crops become unsaleable locally, that is when the crop is not destroyed or confiscated locally by marauders. Where communications exist cash crops are grown for export and the relief of their masters ever burgeoning debt. Costing by the World Bank inform us that with all this help and accompanying baksheesh, the recipients are $100 million per year worse off and the farmers 50 million worse off than if such subsidies were not on offer.

The development of states and spread of globalism in the Third World, means that the subsistence farmer now also needs money. He can work part-time or full-time outside his plot, he can sell his plot, making way for mono-crops if the site and fertility is adequate or he can struggle on. The new beneficial agricultural techniques only benefits those who possess the means of purchase. They need to possess land, monies, have access to credit or influence in the field of politics. Most have none and it is largely the Western World that has furnished an indigenous middle class and confirmed an upper class capable through notional agency arrangements with the means towards such ends.

As the Green Revolution in India progressed, share croppers became landless sometimes by eviction and the rural workforce previously 1/6th landless has reached 1/3rd landless. In Mexico the average farm size increased tenfold leaving 3/4 of the labour force landless and the average number of days worked has halved for them. Even in the U.S., small growers of tomatoes in California have found they cannot afford the tomato harvester and as a result 3,400 growers out of 4,000 have gone out of business in the last 8 years. That is why the consumer has to suffer unbruisable hard skinned tomatoes.

A recent U.N. survey of 83 countries found that 3% of landlords controlled 80% of rural land, and those needing credit not originating from respectable institutions could expect to pay 100%-200% interest.

The I.L.O. but confirms the situation that the greater the production level the greater and more widespread the poverty.

The health food addict concentrates on personal physical and sexual enhancement from exotic additives "untainted" by the disciplines of science, while the vast majority of mankind suffers permanent malnutrition. Man cannot opt out of the fact that 1/2 of the world's crop production is lost to pests, weeds and disease and charity is not even a palliative.

It is too easy to blame monocrops, machines, fertilizers, pesticides and herbicides, especially because they are part of the relatively new and have little history. They have their negative side but in the present state of technology their positive contribution is far greater. And the new is normally more malleable, more able to reduce the injurious aspects. As with oxygen and carbon dioxide they will make a contribution, good or ill, but it is the method of application or concentrations which are the basis of their problem in use, and profit rules here as elsewhere. Erosion has an economic cost element, as indeed has acid rain and pollution of waters involving billions per state.

To repeat the earth already produces enough to feed the world of humanity, were the latter to double, enough could still be produced. The developments considered here can only aid our better understanding and attainment of the best diet possible. It can only help where unforeseen calamity strikes. There are no means of fending off the earths noxious innards or the meteor other than developing our knowledge and this may not be enough.

Mankind in large numbers relies on streams and rivers which from time to time carry disease. In the U.K. we use alum to flocculate and remove pollutant but that costs money. There is a report by M.Kelly on the seeds of a tree at present indigenous to areas such as the Sudan which act in a similar manner and is not scarce. He mentions the Moringa among others already in use as food and decoration. The knowledge of its use in purification needs diffusion, and then application and who knows we could perhaps ourselves gain benefit thereby.

Salination of soil is a problem and 40% of irrigated soil suffers. 10 million hectares are lost annually by rising salinity. As a consequence the introduction and breeding of crops carrying high salt tolerances can become important and success in this area is close. There is a thale cress garden weed tolerant to such conditions which if improved could take salt from the soil. There is a G.M. tomato which can grow in salty water (Indep. 31/7/01). The Saltwort and Glasswort were at one time part of man's diet, certainly they would prove a good food source for animals, and they can be grown on soil containing 30% salt. There are vast areas in Asia and Egypt which suffer deposit of salts from their irrigation systems and without sufficient water to elute it into rivers, and finally seas. With time the water gets trapped and the water table rises. Thus the problem is far from academic. As the table rises, the saltier waters begin to affect the crops and damage can extend more visibly, as has already become apparent on the ancient monuments of Egypt.

Species of bacteria and fungi are being developed suitable for marshlands where the oxygen levels for uptake are low as also on land with deleterious metals and chemicals. Again the osmotic equipment of these sources of food can be successfully tailored given time.

In all this there needs to be the overriding concern with wider and less acceptable consequences, but it is certainly not beyond man to remove or render relatively innocuous any sting in such a tail.

As we replaced the sperm whale as a source for cosmetics and whalebone, the time may well be nigh when the use of animal sources for our medicines, soaps, lubricants may no longer be necessary. There are non-woody fibrous plants, capable of use in the paper industry and renewable, unlike some forests, on an annual crop basis and when properly exploited quite competitive. Cargill Dow , Monsanto and Du Pont are developing biodegradable plastics and fibres from such materials as maize, rape and cress and the genetic revolution has not yet got underway. Many plants rejected in history for out-of-date reasons, are wide open to reclamation as positive contributors to the needs of mankind.

Even were it that the plant itself did not find favour, we now have as an important second string, introduction into their gene structure, allowing their improved exploitation. If as a result we achieve surfeit, then we should perhaps be able to look with a more accommodating eye at that which improves our lives, even were it to cost more. We might become more generous and afford more lebensraum to other forms of life.

Man Remakes The Earth

The concern to preserve our lakes, rivers and forest could be associated with the feeling that THEY (anybody but us) should not cut THEIR forests down, for the same reasons that we in Europe did ,since such action might adversely raise the level of greenhouse gases. There has been talk of the high level of wastage in the clearing and felling of rainforest trees. But, hold a moment "The Economist" of 21/3/98 now tells that 85% of the original Amazon forest is still with us. Satellite photography can be more accurate than experts with their own agendas!

It is now generally appreciated that wood for paper is recycled at an adequate rate but in other reforestation there may be danger in the supplanting of other species by faster growing conifers. There may be a role for Vivitar grass and similar plants, to conserve areas against soil erosion when forest or indeed any land is likely to suffer.

Erosion is no small problem and an excess of fertilizers could hasten the rate. Given no war conditions and the proper application of present day knowledge, Africa could feed and maintain its present day population. We now have good knowledge as the best geometrical arrangements for trees to act as rainbreakers and maintain the soil from erosion lending itself to economies in tree requirements, where this should be a factor.

As in other areas replacing irrational fears by knowledge will allow recognition that there is proliferation as well as paucity available on this earth and we need to know our requirements. Then actions based on back to Methuselah or maintaining the status quo at the behest of voices of secretariats preaching doom, may not find the favour at present accorded them. Past experience and activity will then have its full value in the light of present knowledge. Even the Inca techniques of agriculture now have promise for use in wide areas.

The Development of Resources

Wherever it can be achieved, sustainable development alone is an acceptable option and necessary where it is desired to maintain a status quo. With respect to resources, in spite of voices of doom there is no shortage. Their quotations normally stem from assessed costs at today's figures based on today's technology. If Lord Hanson can raise the value of the reserves of coal in a mine some four-fold to meet an accountancy exigency, one should accept that figures offered related to essential reserves are more concerned with tax and politics, than with accuracy.

In the industrialised world the yearly GDP is $23,000 billion and an average GDP of greater than $20,000 ppa, the remainder including China and India, produce $6,000 billion. However with the more accurate in concept, the PPP, the wealth of the world today is probably nearer $60,000 billion, despite atrocious wastefulness. "The World in 1988" speaks of such a society doubling its fuel use by 2020 and this would presumably add up to a doubling of world wealth or an overall achievement of say $24,000 ppa., but only as an average distributed very unevenly.

These views related to 1986 techniques. They neglected aids from improved fertility, new higher yielding seeds, and the reductions in deterioration in storage and reduction in general spoilage now available. Last but certainly not least there are the prospects in genetic engineering. Between 1950 and 1988 (Econ. 10/6/95) the world population doubled without increased pressure on food supplies. The Green Revolution with high yielding crops increased Asian wheat fivefold from 1961 to 1991. In Latin America steps are now possible to utilise 800,000 sq. miles previously infertile through acidity, by using special but suitable and available crops.

Thus even present techniques can meet that goal. It is all there, ready and available just waiting for the application of this social knowledge so that society can move forward. The stifling impediment, is the present ownership and their controls over men. According to "The Economist" of 22/11/97, S.E. England has a GDP 50% greater than S. Wales. The difference would be greater in Europe, any sort of Europe. Averages never indicate the disparate separation between human beings within the groups, hiding surplus and extremes of poverty alike.

The UN Development Program in the 1997 edition of its "Human Development Report" goes so far as to say that the extreme poverty now suffered by 25% of the world population could be reduced to 4% within 20 years, were there the will, thus supporting the view that Malthus was not always right.

Water

The World is awash with water but only 2.5% is fresh water and more than 1.5% of that is trapped in Antarctica and Greenland. Yet there are few problems in relation to water supply that cannot be solved. Within the bounds of present technology there is no need for water scarcity in Britain or indeed N America. Scotland could meet any foreseeable additional needs and Canada has 20% of the worlds fresh water supply, most of it in the Great Lakes.

The problem lies in conflicting control and costs and the reconciliation with values.

However one billion human beings lacks access to clean drinking water! Preventable water borne disease kills millions and discomforts many more. To eliminate disease from waters and give adequate sanitation in the world would need some $2 billion and the savings in disease and loss of ability to work would amount to some $16 billion (Economist 15/5/04),yet we are slow to move, perhaps because it is a cost to the West and a benefit elsewhere. There is also a cause for concern that 40% of world food is raised by irrigation when much of such water could come from recycled sources especially as this would stay soil fertility.

Elsewhere it need hardly be a great problem but for the reason of state control. As a result tensions are ever present between Ethiopia, Sudan and Egypt for water and Turkey, Syria, Iraq, Jordan and Israel as to sharing their joint sources of water.

However, too much is wasted. In Great Britain as much as 30% of domestic water leaks away before use, and there is growing concern at using drinking water to flush toilets, spraying gardens and golf courses and even to spray non-food crops. In California 160 billion gallons of reclaimed water per year is so used. The modern flush uses 1.6 gallons the older use 5 gallons and at that level were considered to take up 30% of domestic requirements. There is also no absolute necessity apart from slightly increased costs and social attitudes to treatment upgrading water to drinking standards where desirable. On a global scale that is how we receive it, Holistically it is already the done thing.

Before 1939 over 60 tons of water were required to process one ton of steel, today the figure is 6 tons. In Japanese production there was a fall of 30% between 1965 and 1989 in water usage, and in the U.S. today it is 20% lower than in 1980. But we must also be aware that in a factory producing chips for semi-conductor industry, ie a new product, one day's production entails 3 million litres of water and safe recycling is very much a necessity because of the demand by the process on purity.

There has been great progress in the use of osmosis to reduce the cost of desalinating salt water to 50 cents/cubic metre in the U.S.("Economist 4/4/98). Drip systems as used in Israel can save 35%

The trial use of polymers to hold moisture falling on land in Egypt has led to the finding that the presented super absorbent materials when present at 0.2% in the soil reduced the water requirement for a particular crop by half. Drip systems are a little more expensive than full spraying but have saved 70% in India, Israel and the U.S.35% of water and low energy sprinklers are nearly as good. Irrigation allows 2 or 3 harvests and these new techniques are far and away more able to improve yields above channelling. Since they use less water they reduce problems such as waterlogging, salination, and erosion significantly.

Certainly in G.B. there is no shortage of water falling on any area of land if we were willing to use and pay for the means available to us and that expense would not be great. Yet every year there seems to be a water crisis! Talk is then about metering, using canals, piping from far away places and dams. The facts are that the failing infra-structure allows more than 25% of good water down the drain, and why do we never hear about the enormous quantities of water wasted by industry where it would be economical to recycle? Evidently metering has not encouraged them. It may be necessary at some time to remind people that consumption of water is only 10% of the cost. It is just another example where talk of freedom is channelled into controls, helpful to meter manufacturers but of little use in solving the real problem. We would remark that (Indep. 14/8/97) there has arisen a strong cry over the damage that could arise as a result of the ground water level rises in some major towns. In London the rise is some 3 metres p.a. and boreholes are now being considered (F.T. 24/5/98) to prevent damage to foundations and basements.

Recycling by Industry has already reduced their requirement by 1/3rd from their level in 1950. Water for farm irrigation has dropped by 11% by improved techniques and even the domestic user has not seen a rise by the better utilisation of water in domestic equipment and short-flush toilets.

Returning to the theme of water supplies and the role played by those magnificent dams. We have already dwelt on their displacement of millions of little people. People previously "comfortably" settled in fertile areas, are often given promises of compensation which are derisory or which soon become too inconvenient to meet. These are destined to the dustbin: they are meant to become economic refugees with all the malodour normally attached to this category even within their own state. "The Economist" of 23/12/95 estimates that 75% of monies invested in large Third World public sectors has been pocketed by officials, or in construction frauds or just lost. There is normally destruction of forest land and erosion can follow. The returns are not to the poor and there is consequent eruption and growth of deleterious insect populations as evidenced by the rising bilharzia rate.

The dams do however offer electricity and the availability of water for irrigation. As a result irrigated farmland has increased threefold in 40 years and grows one third of the world's food. However possible long term salination could decrease the level of fertility.

The draining of the Wetlands also has its price and here 56 of 152 wetlands have been damaged by water abstraction.(Indep. 11/6/96). Wild life is decimated and any alteration to the water table contributes to salination, but this decimation also diminishes the area covering the activity of the mosquito.

Dams have decimated river fish life in Russia as well as destroyed by displacement the marshes with all the attendant wild life therein. Stemming the flow of water in Spain threatens rice production by allowing the entry of waters from the salt Mediterranean. Even that glory of the State of Egypt, the Aswan Dam stems nutrient supplies downstream. Where it enters the sea, the fish catch has been reduced to 1/6th of its former self, but there is little ability to complain. (Econ. 23/5/98). In India there are some 3,600 dams which have displaced 150 million mostly to the shanty towns according to a letter in the" Independent" of 1/11/99.

Nitrates and phosphates have had attention as water contaminants, but there is no doubt of the vast amounts of treated and untreated sewage as well as the many chemicals which enter the ocean beds. Treated does not mean total removal of all possible problems. Treated is a legalistic term. Success is dependent on care and site but we are bedevilled by pressure groups concerned to win a case rather than supply the fullest information.

Waste disposal

Concentration in cities has led to corresponding levels of waste. Social standards and attitudes, followed by knowledge have given us an awareness that waste materials, especially malodorous materials in concentration, could become a serious source of disease. Disposal as far as possible from the person has become a matter for urgent concern. The ideal sinks were such places as the ocean and there was consensus, at least until the 1930's, that this was the ideal limitless sump. It was not appreciated that the North Sea was largely less than 100 feet deep. Better understanding of waste products including effluent, coupled with recognition of the much increased volumes and range of chemicals entering this arena has led to a more cautious approach.

It is increasingly obvious that effluent now enters the North Sea in enormous quantities, including huge arrays of chemicals, and some even at trace levels might produce unacceptable results, even constitute a danger. Of thousands of different chemicals released from G.B., only 16 are monitored, and they may not serve as adequate markers for others. Deep sea dispersal using depths below 3,000 metres has also been mooted, since there is little life at this level, and 1/2 the oceans offer such depths. Land fills are used but are filling up quickly. Increasing legislative control is concerned with seepage into drinking water etc, and this means extra costs. The cost of landfills for hazardous waste has increased in ten years by one hundred times, and the increases for more acceptable residues have increased between 4-40 times. Transport looking for landfill now needs to go much further afield and this again adds to cost. Of course in the end a completed landfill is a candidate for building but on average, it takes at least 20 years before it is fit for such a purpose and may need deep piling. Meantime decomposition by bacteria etc gives rise not only to carbon dioxide but methane and were it to be usefully organised, not only would the danger of ignition disappear but it could be channelled into supplies of heat for the local communities. One hears that quite a few landfills are capable of heating 10,000 homes each.

Research into the conversion of cellulose from household and agricultural waste to produce fuel is near the point of market prices and this alone could deal with half our waste.

The Green NIMBYS, object to the import of 80,000 tons per year to be rendered innocuous by incineration. While this amount could grow, it should be borne in mind how small this amount is against the nearly 4 million tons arising in this country for such disposal. 300,000 tons are imported for land fills. The real problem is that most waste disposal authorities here ,have no adequate means of monitoring such operations, and standards are very "variable".

Landfilling in 1988 G.B. took 80% of waste, incineration 2%, the remainder being treated or dispersed at sea. High temperature furnaces exist for reducing or eliminating dangerous products and any consequent processed products, but these too meet with much Luddite opposition.

Waste still grows at approximately 5% per year. There are already fears leading to the always desirable end of recycling, reconstitution and designing towards that end. Some perspective is inserted (Econ. 29/5/93) if we take account that if U.S. waste continued at present rates for 1,000 years and all put in a hole 100 yards deep it would still occupy 90 sq. miles in a country of 3 million sq. miles. Remember too that waste does not just appear at disposal sites. It needs to be transferred by men and equipment and automobiles and costs much energy in all areas. There is therefore a need to add relevant collection and disposal costs against costs involving incineration or recycling. There is also the cost which never seems to enter accounts of dealing with contaminated land, but here as already brought to attention there are plants and bacteria to be applied for such a clean-up.

On average 2.5 tons of selected waste could produce 1 ton equivalent of coal in energy. 50% of our domestic waste is paper based, 40% includes metals, glass, food and plastics. The remainder is rubber, leather and textiles. In all this the important element, sorting, receives least attention. Possibly at least 75% of our waste is recyclable at least once, and this takes no account of the possibility of redesigning other articles so that they can enter that magic circle. Using the incineration route in G.B. could convert waste to 8 million tons of coal equivalent per year and it may not be that expensive in relation to the developing picture of a barrel of oil costing 40$ against present day $60.

Sewage plants could concentrate their methane and use it as a fuel and there have been successful attempts to use selected waste with solid fuel. In 1986 it was estimated that possible power from this U.K. source was equivalent to a million tons of coal per year. Instead of energy utilised green gases we have unutilised green gases.

Some waste is due to designed short-term obsolescence, which is socially unacceptable except in special cases. It must also be obvious that the current practice where all waste is mixed and not separated is a barrier against best extraction. Sometimes incineration can prove a better mode than recycling, using landfill or the ocean as a sink, and this must take into account the pollution effect. There is no doubt that incineration and where called for high temperature incineration offers better control than the other sources mentioned since scrubbing and neutralising equipment can readily be included.

If one can use energy from the combustion of waste products then the economics may in certain cases direct one into that area and in Scandinavia much use is made of this resource with Germany not too far behind. There is however the case where high temperature incineration is the only course especially where the product would otherwise spell danger.

Technology is at a phase where bacilli are being developed to extract metals and even poisonous chemicals from waste.

The feared toxic effect of dioxins and furanes arising from incineration are not of themselves unreasonable. It is however pertinent that these emissions have arisen ever since the first forest fires of pre-history, and where relevant, high temperatures above 850 C is adequate for their removal. The inhalation level locally is but one per cent of that already present in mother's milk in normal cause and in dioxins we are talking of parts per trillion. In 2005, Yuschenko was poisoned by a dose 6,000 times the normal level of a dioxin used in the Viet-Nam war. He has had treatment and is actively alive to date.

An incineration plant for a city of 200,000 could cost £20 million.

The dioxins are a man-made group of chemicals and therefore the Greens need to consider it an Evil. There was Seveso, and epidemics to whales, seals, the ozone layer and striking closer to home, fertility. One would think that the sperm count was down to a hundred rather than to the multi-millions still available. We have this scare mongering in the" Independent" of 25/11/97 telling us in one article that the sperm count for American humans is half what it was in 1938, a mere 50 million per shot, 23 million in Europe, yet 100 million in New York, and the cause automatically given as pollution defined in "The Green ways". Apparently the U.S. and especially New York ares doing the right thing sometimes. Such offerings especially where there is no evidence given as to any greater deterioration or malformation are more to do with the prevalent agenda to criticise man's activities, rather than more objective concerns. Now there is mother's milk and we must agree that at present high levels therein this subject needs concerned and immediate attention, since it could affect the viability of Man, but we could do without unsupported claims of imminent disaster. We have even had support in 1997, for the issue of non-PVC credit cards on account of phthalate and dioxin content in the present versions when they are not even present.

If the cost of recycling contributes a greater value than incineration then this should be a preferred route but the nature of the fuel to be utilised must also be a consideration. If one allows that the Western World produces one half ton of waste per year per person and the Third World a quarter of this figure, the huge nature of the problem unfolds. American domestic waste includes yearly, 1.6 billion biros and two billion razor blades. Only some 10% of waste is recycled. P.V.C. polystyrene and polyethylene etc used as packaging products also play their part. Waste figures show paper as 40%, glass at 10% and there is organic waste. Correct sorting helps and even more important is the availability and proximity of efficient processing plants. Since the potential value of this raw material is low, the cost of transport must be kept low and as a result collection or treatment centres cannot be concentrated into large units, even were that to offer seeming economies. Smaller units also need more staff and more skilled staff per unit production, but the developing shortage of landfill sites and resultant higher charges, does move the goalposts.

Packaging does not help to save. It organises the availability incorporated in one package of far more items than one needs or is ever likely to use. You seek two screws and are offered a packet of ten and you seldom use those remaining. Some part of packaging is protective or purposed for efficient transport but a great proportion is designed to attract the buyer to the product. One sixth of a shopping bill can arise from packaging.

The problem which has made headlines recently is of large vessels ploughing the seas of the world for an opportunity to unload unwelcome refuse. This includes poisonous chemicals often with identity tabs deliberately removed. There is inadequate provision of world capacity to treat these products and as a result any values in them cannot be recovered. Third World States are the interested candidates for such residues, where they remain dangerous, especially when with time the containers corrode. There is also practised incineration at sea, which not only wastes energy inherent in the material, it wastes energy in transport, but easily avoids supervision.

Western obsolete shipping is sent to break-up yards in India and Pakistan where the legal requirements for working with dangerous materials or practice are less costly and less safe. China is also arising as such a venue. Here (Times 13/11/03)there is outcry not at this practice with its shadow of Bhopal, but at Hartlepool , U.K. becoming a site despite the fact that there sufficient safeguards have been incorporated meeting the Environment agency criteria ie adequate from a safety standpoint.

Metals are a major product for society and often difficult to produce. Estimates of losses from corrosion are some 3% of G.N.P., probably a considerable underestimate. Yet at least 20% of this vast figure, comparable to our third world grants, could be saved merely by the application of well-known electro-chemical processes.

The contamination due to the growing mountains of rejected waste materials could with little effort be reclaimed for use at the previous level or at lower but still useful levels. Attending to this need of society is beginning, but only just beginning.

Acid Rain

Acid rain has been under examination and relates especially to the rivers, trees, life in rivers and not least important the very air we breathe. The main culprits accused, and as per practice condemned without much trial on circumstantial evidence are sources, giving rise to sulphur acids, and nitrogen acids. Twenty years ago (Independent 4/9/97) the demise of vast German forests was seen as inevitable, now we gather that the condition of the trees has been improving since 1994 and the proportion of severely damaged trees is now 20%. No specific cause or causes have been isolated, but it could be correlated with the demise of industry in Eastern Europe which used coal of high sulphur content. There is also the diminution of pollution from the car.

Most important is the fact that it was not that difficult to reverse. A look at attributed sources, shows much more concern with man-made sources than others, and less concern with its importance. "The Sunday Times" of 13/11/05, tells that wild life in the affected areas, including brown trout ,caddis larvae, other insect life, algae ,mosses and a number of plants have all returned. Professor Battersby of University College has found acidity halving in less than 20 years, and the lower use of sulphurous and nitric oxide acids will have made that contribution as well as anti-pollution devices.

Much encouragement including tax advantages has been given for maintaining forest and reforestation, then comes a tax change in 1989 G.B. and 40 million young conifers are burnt down, without concern. Even were permanency assured the perceptions of today including acid rain, raise doubts as to the value of the contribution made.

Acid rain has contributed to the deterioration in river life and is said to be exacerbated by planting of conifers. The increase in the shading of rivers by trees is said to adversely diminish the heat radiated by the sun and affecting temperatures. Further it has been suggested that in upland forests increasing the tree population, increases the acid pollution by acting as a condensing medium and the acid makes its way into streams and lakes. Thus such problems can only be dealt with by global organisation .Again "Independent" 13/3/03 there is some evidence that pine trees under certain sunlight conditions can release smog-making nitrogen oxides. It is but a further reminder that the problems and solutions put forward are seldom as simple as made out.

There are American figures indicating that reducing sulphur dioxide may help, but there is the recent appreciation of the role of this gas when present in clouds as a reflector of heat from earth. The volcanic eruption in 1883 at Krakatoa resulted in a dip of world temperature by 0.2 degrees centigrade for over 2 years, It happened similarly in Bali in 1963 and of course there are many others such events, occurring at different levels of emissions. The sizeable reductions in the emission of sulphur gases from electrical generators in G.B. and other places at enormous expense may therefore not prove that important.

Attempts at reducing the acidity levels of lakes with calcium carbonate and the like have not been altogether successful in that the new milieu arising is sufficiently different from the old and the old status quo not quite restored. While fish have returned certain lower forms which are sensitive to calcium have not. There is the growth of "unwanted" quantities of algae on various waters and we read of the trial use of the water flea to control the level so that it may enable a return to former conditions to take place. However this is but an attempt to avoid sewage stripping techniques already in successful use by concerned countries and at minute cost. If care is not taken it may go too far for the maintenance of the living creatures of interest to ourselves. Already there are hints that the cleaning up of the River Trent has diminished the coarse fish therein because under the recently attained levels of hygiene there is not sufficient food.(Independent 4/5/99).

Again the increased concentration of nitrates in the North Sea has caused a large increase in the plankton concentration and this is a base material of the life cycle.

Volcanoes throw up enormous quantities of sulphur. Mount Pinatubo erupted in 1991 sending at least 15 million tons of sulphur dioxide into the air, certainly ten times man's yearly production. Pinatobo was certainly not the largest of eruptions in the last 100 years. Again plankton in the North Sea is now considered as contributing 1/3rd of the acid rain of Scandinavia, previously attributed to British coal used in electricity generation. Recently it has been claimed that N. America is also making a Canadian contribution.

There is much plankton in the rest of the world producing methyl sulphides, which gives rise to atmospheric sulphur gases and plankton and its bye-products have been around a long time. In this we do accept the point that man-made products can be more easily undone, but a prime requirement is data on the scale of contribution from each source.

Ozone in the lower 10 kms of the atmosphere tends to decompose and by further reactions with sulphur and nitrogen oxides form sulphuric acid and nitric acid. These acids can travel over a wide area in clouds and when precipitated can affect micro-organisms and more. The role assessed for these acids in forest deterioration is not clear since we are not sure whether they weaken aggregation, weaken the root structures, leach essential element of nutrition away from the trees or by reactions release poisonous elements. It is put forward that many temperate forests are in decline. Plants in the Cairngorms are said to be damaged by acid mists and melting snow , a worse hazard. It is claimed that melting snow when reaching streams give 3 times the pollution of that where there is no snow or ice and spring is a melting period when life is also at its most tender.

Acid rain has been in the air for a very long time. In 1858 it was already a subject of concern and attributed to sulphur from coal. From ancient times, creations at the bottom of seas existed with a life cycle tied to sulphur, not oxygen. Where recognised as in N. Canada and Greenland these have given sulphur oxides over an area of 25 square miles and for aeons of time.

Terrestrial and aquatic life can still flourish in rainfall of a very acid nature. It is therefore not just the acidic nature per se that needs investigation but also its secondary effect on releasing or removing essential elements affecting metabolisms. Many lakes in the northern hemisphere are crystal clear because of acidity and the consequent absence of bacteria and photosynthetic plants, but others are covered with moss and algae. life can therefore survive these conditions and knowledge may be desirable to ensure that it encompasses life of interest to ourselves.

The choice of acid rain as tree killer is almost without apology gradually but positively being replaced by ozone arising from car exhausts. There is the effect on streams to be considered on a world-wide basis but this may only be serious when it is held by snow and then suddenly released in a spring melting.

If nitric oxides, sulphur oxides, lead, carbon monoxide and ozone arising from man are important causes then the developments of the last few years could see their end, especially in automobiles. Engelhard have already developed catalysts for destroying ozone using the huge surface of radiators. Coal generators can already have sulphur oxides removed (at expense) and in this field the use of cyanuric acid offers at a cost the removal of nitrogen gases. In the latter course there would seem to be much cheaper alternatives to cyanuric acid in train. Here "Greenies" have played an important and useful part in a desirable clean-up. With respect to trees Dr. J. Innes expressed the view in 1990 that for trees there is as yet overall, no real decline in tree health and no link to be seen with air pollution effects.

The recent considerable effusion of chemicals from Switzerland via the Rhine, killed an estimated 250,000 eels and vastly affected existing fish and river plant life. Death could be seen a 100 miles from Basel and the river was full of pesticides and fungicides, yet 10 years later it has recovered. All has been transported to the sea. This was a one-time happening, but end-products are accumulating by regular dumping into the North Sea and as to its transformation, silence rules.

It would undoubtedly be wise to delimit this amount even were no unhappy consequences foreseen, the scare mongering tactics with respect to seals etc notwithstanding. It really is not good enough that England only tests some 16 of the many hundreds of residue compounds that find their place in the North Sea. If events became out of hand then the consequences could not be rectified quickly especially were they to enter the human food chain .

Life is astonishingly flexible in its ability to cope but there are limits, some of which we do not yet comprehend. One should then act on the basis of the worst scenario within current information, especially where an alternative path is economically unavailable or the benefits small.

The ground has now shifted to include these with D.D.T and other insecticides in the hypothesis that they can play mimicking roles to hormones and interfere with endocrine reactions in the body. What is not mentioned is that much of plant life contains compounds equally capable of such mimicry.

What the pro-animal lobby does not seem to have learned is that while animals such as rats can be used in attempts to parallel man in their reaction, man is not a rat. In like fashion seals and dolphins may not suffer. The animal lobby which always tells us that results on rats are totally inapplicable to man, should accept such where the results do not speak in their favour. Animal deaths may have had other causes, not the simplistic unproven umbrella evil, which in this case replaces God as an explanation and again, does not require the rigour of examination. The seal from 1998 thrives, and there is no longer the talk of pollution decimating them but of a virus which returns intermittently.

The public consensus established by the various lobbies and its acceptance by politicians means indulging some of the most irrational fears as well as real danger for the majority of the populace.

As a result the expensive decisions related to sulphur dioxide and the dioxins may lead to the betterment in removal levels desired, but not necessarily to any useful improvement in man's condition. Today (Economist 21/9/02), modern systems can remove 95% of sulphur and more cheaply than before. There is the synthesis of methanol from coal derived gas. There is now in late development the IGCC form of generation which converts coal to a gas which is cleaned and burned in a turbine. The result is that carbon dioxide is extracted more efficiently as is sulphur dioxide, tars and oils. It can also be fitted to older designs of plants. This new design should offer a 40% conversion against the present 33% and could reach 44% i.e. 20%, possibly 30% savings in energy.

The net service these bodies perform should not lightly destine the poor among us to the consequences of pollution by the NYMBY approach of the upper classes. Disraeli organised the sewers of London, but his major concern arose from a desire to avoid the perpetual stench that M.P's suffered at Westminster and the increasing awareness that the arising diseases could spread from the poor to the elect of our society. More recently when a Japanese firm released mercury compounds into the local estuary and when Mexican industry released poisons into the river in Matamoros feeding Texas, more than the locals or the poor come within the orbit of the endangered.

Mexican industrial plants with lax rules on safety and pollution, are often subsidiaries of American groups attracted by these considerations as well as cheap labour. Their polluted effluent often untreated flows through the towns that have grown around them, in open ditches but deaths in this land are not as closely linked with cause as in our world.

Chemicals

There is still much hysteria relating to the use of insecticides, such as D.D.T. banned for use in the U.S. since 1972, and Aldrin also banned and largely for political reasons (Independent 5/10/94). Any serious consideration except on a black and white basis must lead to the conclusion that its use has been a major good. It wiped out malaria in Greece, Italy and the S. Americas. Today in Africa south of the Sahara 2 million die annually, from malaria, sleeping sickness, typhus and yellow fever, yet the supposed deleterious nature of D.D.T. is not that clear. In 1971 the Environmental Protection Agency declared that it was not carcinogenic, mutagenic or a hazard to man. They also declared that it was not deleterious to fish, wild birds. These findings were kicked into the long grass by E.P.A. on the ground of caution and no further evidence against has since appeared Here too the" Independent" 28/5/93 tells of bacteria capable of rendering it innocuous, were that necessary. In 20 years of use, some 500 million lives were saved. The W.W.F., no friend of chemical manufacturers, has tested 41 synthetic chemicals including D.D.T, P.C.B's., polybrominated phenols without finding adverse positive evidence. Lindane too has done more good than harm although considered as a potential carcinogen. True, information has appeared that both can enter the food chain and remain stored for a long period in the fatty areas of mammals. True, that alternatives now exist which can be used in lower concentrations and have a lower active life, but in the Third World price is the factor which often decides against the use of medicines and protection. In that area human life is valuable to few of their controllers, and little money is available to pursue such avenues. It may be that the developing price for newer products such as Icon may equal that of D.D.T. but then there is the cost of changed techniques and training. The main justification for the newer chemicals is the necessity to introduce changes when immunity to the prevalent material becomes manifest. Nobody has yet shown convincingly that the long life of D.D.T. has affected our health or lifespan. Again as with the case of dioxins and other chemicals the effect on bird life etc must bear consideration, but should not be the arbiter. The increasing number of insect species showing resistance and more to its toxic effect must remain the important consideration. Historically it should be noted that its use in the post war typhus epidemic, and for malaria has improved the lives and lowered the death toll for billions of human beings.

That fact remains its justification for further use. Since we have wealth enough a better society would be concerned with best treatment not lowest price. Behind much of the argument is a distortion of science, against man's intervention to secure his lot by chemicals absent in nature.

Seveso, where a chemical plant exploded "showering" a wide area with dioxins gave rise to the most fearsome of descriptions which at the time were not unreasonable. It was put forward that the area could not safely grow food and that birth defects, chromosome damage and cancer was to be expected, and indeed these were possible. The problem ranged more widely since in the manufacture of many herbicides and their distribution higher concentrations could arise and even in general dispersal, dangers could surface. Wood and paper so treated could be used in disposable nappies as well as in electric transformers. However of 210 dioxins only 17 are considered as possibly toxic, yet all have entered into the mythology as dangerous.

In Monsanto trials commenced in 1948 under manufacturing conditions, nothing positive was revealed apart from some tobacco plant acne. In Sweden some forest sprayers have suffered some 6 times the sarcoma effect present generally in the population, and the concentrations to which they could be exposed could be high since supervision in use could be low. Other studies have given negative results.

It is of course very difficult to forecast long term results. Hamsters were found very sensitive and man insensitive to such materials. Chloracne is in no way a welcome contribution, but it is not a danger to viability. The Seveso episode to date, has indicated that even humans receiving large doses have not suffered death nor any of the major ills forecast. It also confirms that we are neither hamsters nor monkeys. In 1997 it has been mooted that they may affect sperm production. Providing quality is retained a reduction from 100 million in 1940 to "only" 66 million in 1990 is not quite catastrophic.

In the soil P.C.B. has a 1/2 life of about 10 years, in darkness 3 years, but in sunlight the period can be measured in days. It is present in mother's milk and can accumulate to an equilibrium level in man but at that level has shown no sign of danger. Birds seem able to absorb considerable amounts and some seabirds have as much as 1000 parts per million. With respect to seals and the like, it is hard to believe that this compound is the main causative factor of death or epidemics, indeed comes 1998 and the claim sinks underground. They were subject to these long before man produced these compounds.

Just in case matters turn out worse than has been indicated here, hard working scientists, not Green hot gospellers, have been busy developing bacteria to degrade such product and render them safe. They found this class of material was attacked in Hudson river deposits by bacteria and are developing this information. A G.M. tobacco plant has according to the" Times "of 1/3/05 a part to play in decomposing P.C.B's, and Dioxins.

The Great Lakes and St, Laurence between Canada and the U.S. have benefited from the lower levels of steel production etc and heavy metal residues in fish are much reduced in 2001. (Econ 17/11/01) They still suffer from the huge quantities of manure spread on farms.

There are the popular recent claims, based largely on flawed experimentation that chemicals including plastics have reduced the sperm count from 113 million per ml to a mere 66 million and as affecting sexual organs. We have also had similar alarms of disease caused by mutant fish , disposable nappies, food containers cosmetics and anything made by man. With respect to dioxins (INDEPENDENT. 29/6/01) The dioxins released from municipal incinerators is 17% of that released on a Guy Fawkes bonfire night and wherever fires from wood or grasses or cows or sheep etc dioxins are produced.

Lead is on its way out as a pollutant, but even here our new understanding from astronaut activity has begun to confirm why hot bath s seemed to relieve lead poisoning. As an alternative route bacteria and plants are being developed capable of accumulating 80% of their bodyweight in metals. Some can absorb 90% of radio-active metals and could clear an accidental seepage over their lifetime. In the Netherlands a pilot scheme shows that mercury at 2ppm can now be reduced to 5 ppb. The plants of interest are part of the alyssum family and it may be possible to develop them as vegetation ores. To date manufacturers find it cheaper to pay fines than invest in such treatments. (Independent 17/1/94). The Economist 23/11/02, tells of the use of bacteria to extract and reduce copper and nickel sulphide ores to metal oxide bases without high temperatures and the emission of sulphur compounds and also at 2/3rds the cost for metal production. These are promising avenues for the advancing of quality of life, developed by science not by the Luddite Greens.

The Greens having discovered that their once lauded unleaded higher grade petrol is more damaging than the 4* grade it was meant to replace are letting this view to die very very sotto voce. The benzene in the new mix can contribute to leukaemia. Their vociferous claims that air pollution is a major cause of asthma are far from proven according to work quoted by the "Independent" 20/9/25 and their stress on that as cause has meant the side-tracking of investigation into other areas. A finding was that there was the same level of asthma sufferers in the Isle of Skye as elsewhere. Now we find that introducing the car catalyst converter increases the use of fuel, increases the carbon dioxide fed to the atmosphere but decreases the carbon monoxide.

"Scare Stories" BBC2 of 11/12/97 makes quite clear from Dr. Johnson ,charged with Greenpeace scientific research, that FACT was to subserve a political agenda. Where, as with Brent Spar the ecological merits countered their view this was of no moment. They throw up phthalate plasticisers as a cause of cancer and male sterility. 1998 publications of work by the European Scientific Committee on toxicity etc revised statements on cancer since the mechanism for rat cancer is limited to rodents and we are not rats. With respect to the plasticisers in commercial use none have been shown to shrink testicles, but such facts do not enthuse the media. They throw up so many thunder clouds that one cannot distinguish those that bode harm. They are concerned not to observe that many plants mimic or contain hormonal substances that may equally be deleterious in use. As to government secrecy, the Dept. of Environment by refusing investigation at the relevant Atlantic site originally destined for Brent Spar, ended that chance of evaluating pollution of the seas. Freedom of information could not have helped, nor direction of research grants or perceived career threats.

Actions to make better use of power from the tides etc. have hitherto, inevitably led to changes in the local equilibrium of nature, requiring closer examination of alternatives and how far they may need amendment. Especially important are changing concentrations affecting the flow of sea currents with their major effect on life in the sea and the climate. A new tidal design is now in consideration to serve Wales giving the normal ebb and flow mechanism and able to supply 14% of its needs and there is the possibility that using wind generators on that site could double this amount. We are maximising the potentialities available but there is the drag back of the needs for capital.

Today the politicians are disadvantaged with respect to the greenhouse lobby and its easily assimilated verbiage. They bend their knee but their drive as motivated by the logic of capitalism is to block, frustrate or slow any movement. Existing facts need untangling, important facts need to be researched, and political colours should not be allowed to intrude or obtrude.

The media, the "health" addicts and the "Green's" tell us that 70% of the remedies for cancer and other ailments reside in the rain forest, just waiting to be picked. Always there is the approach that the synthesised medicines of today, with their separation of the contributory ingredient and rejection of the innocuous or deleterious is essentially sub-standard.

The reason modern medicine has largely displaced herbal remedies is simply that the latter has not been successful in curing a large swathe of diseases, especially those that are malignant. Many are largely untested, depending on anecdotal evidence. Most herbal remedies come under the food umbrella which does not requires the efficacy and stringent testing required for today's medicine. Of course untested does not of itself mean useless.

They never inform us that the exercise of examining the enormous number of plants and insects for use as medicines may have parallels with seeking needles in haystacks, except that the value of the prize is much greater. In this endeavour, one is seldom advised to seek prospective remedies on the M25 or M1 and one could be wrong. The forest as a source of the unknown could hold unknown remedies as by definition of the word unknown. The public is not made aware that the vast majority of medicines are derived from plants and amended as in aspirin only to eliminate their bad side effects.

Again and again we see people dressed in their comfortable clothes, perusing esoteric labels in health shops and not aware that the main problem in the world today for 90% of the world population is not the C.F.C.s, not the carbon dioxide destroying our habitat but the overwhelming, everpressing and immediate necessity to procure the basic food and shelter for themselves and their families. The third world is not as yet concerned that in ignorance, their means of sustenance is depleting the earth's resources and may cause an Armageddon. Such fears are not for them when they can see the searing flames of disaster and starvation about them throughout their lives.

The hankering after vegetable life, and ostensibly animal love, can only lead to the near elimination of domesticated animals serving as food and their destruction as a valuable food source. It leads also to the efforts which releases mink to become a pest to the inmates of the woodlands. It releases experimental poultry to be eaten alive by predators. It is not love of animals that is here displayed, but hatred of man which is expressed in such actions. Every medical treatment, operation, blood transfusion can be, and is by some, construed as interference with nature or playing God. The throat operation on the Pope in 2005 comes into this category. 85% of animals in tests are rodents, and 3% dogs cats monkeys and larger mammals. Dogs are 0.4% Monkeys 0.2%, and it is important to realise that monkeys and dogs are too expensive for wider use. It is money not ethics that limits the use of animals to the minimum. While the number treated per year in the U.K. are 2.75 million the great majority suffer minor distress or pain, alleviated by pain killers or anaesthetics. In the 1970's it was double that figure. Where necessary they are killed without pain and only 2% suffer severe pain. 55,000 is a large number but the suffering of man should be of more concern. (The Economist 27/8/05).

To repeat nobody uses animals, especially dogs and monkeys where other methods are available because they are very, very expensive. Microsurgery for example requires careful control of bleeding and only live animals can supply the necessary data.

There are 40,000 chemicals in use which have not been tested for deleterious effects and where the choice is between animals and man for the author there should be no contest only regret.

The vast majority, 40%, are used to test new treatments and 30% for fundamental research. The list of successes is long and forgotten by the public, yet most important. The success of medicine using vaccines and antibiotics for polio, tuberculosis, diphtheria, and vaccines, antibiotics, anaesthetics, transplants, blood transfusion, insulin, asthma, cancer, hepatitis even high blood pressure and HIV has to be acknowledged. There is smallpox which was devastating in its effect now eliminated. Not so long ago tuberculosis claimed 2 per 1000 now it 1 in 100,000. To this list let us add kidney dialysis, by-pass surgery and radiation studies. The work of Blakemore has contributed to the treatment of child blindness and also Huntingdon's disease. Work done on 7 monkeys in 1990 at Radcliffe Infirmary has allowed 30,000 Parkinson patients to benefit from implants. Any of these introductions are more carefully tested and monitored than foods such as nuts. Where possible tissue cultures are used and pain avoided. One needs perhaps to add that the O.E.C.D in 2000 A.D. introduced measures to reduce the number of experiments on animals by 17%, equivalent to half a million in Britain. However animal testing is still necessary for disease research on vaccines and Alzheimer, multiple sclerosis and cystic fibrosis. Everything we have learned about the brain and its functions has required it. In Cystic fibrosis , mice are created with the same genetic effect and success would save many human lives. There is even a glimmer of hope in that a breed of mice has been formed which can regenerate heart, toes, joints and tails and cells from such mice can be passed to other mice for help in their generation. (Sunday Times 28/8/05). How could the hope of amputees for regeneration be sustained other than by such experiments. `

The hankering after vegetable life, and ostensibly animal love, can only lead to the near elimination of domesticated animals serving as food and their destruction as a valuable food source. It leads also to the efforts which releases mink to become a pest to the inmates of the woodlands. It releases experimental poultry to be eaten alive by predators. It is not love of animals that is here displayed, but hatred of man which is expressed in such actions. Every medical treatment, operation, blood transfusion can be, and is by some, construed as interference with nature or playing God. The throat operation on the Pope in 2005 comes into this category. 85% of animals in tests are rodents, and 3% dogs cats monkeys and larger mammals. Dogs are 0.4% Monkeys 0.2%, and it is important to realise that monkeys and dogs are too expensive for wider use. It is money not ethics that limits the use of animals to the minimum. While the number treated per year in the U.K. are 2.75 million the great majority suffer minor distress or pain, alleviated by pain killers or anaesthetics. In the 1970's it was double that figure. Where necessary they are killed without pain and only 2% suffer severe pain. 55,000 is a large number but the suffering of man should be of more concern. (The Economist 27/8/05).

To repeat nobody uses animals, especially dogs and monkeys where other methods are available because they are very, very expensive. Microsurgery for example requires careful control of bleeding and only live animals can supply the necessary data.

There are 40,000 chemicals in use which have not been tested for deleterious effects and where the choice is between animals and man for the author there should be no contest only regret.

The vast majority, 40%, are used to test new treatments and 30% for fundamental research. The list of successes is long and forgotten by the public, yet most important. The success of medicine using vaccines and antibiotics for polio, tuberculosis, diphtheria, and vaccines, antibiotics, anaesthetics, transplants, blood transfusion, insulin, asthma, cancer, hepatitis even high blood pressure and HIV has to be acknowledged. There is smallpox which was devastating in its effect now eliminated. Not so long ago tuberculosis claimed 2 per 1000 now it is 1 in 100,000. To this list let us add kidney dialysis, by-pass surgery and radiation studies. The work of Blakemore has contributed to the treatment of child blindness and also Huntingdon's disease. Work done on 7 monkeys in 1990 at Radcliffe Infirmary has allowed 30,000 Parkinson patients to benefit from implants. Any of these introductions are more carefully tested and monitored than foods such as nuts. Where possible tissue cultures are used and pain avoided. One needs perhaps to add that the O.E.C.D in 2000 A.D. introduced measures to reduce the number of experiments on animals by 17%, equivalent to half a million in Britain. However animal testing is still necessary for disease research on vaccines and Alzheimer, multiple sclerosis and cystic fibrosis. Everything we have learned about the brain and its functions has required it. In Cystic fibrosis , mice are created with the same genetic effect and success would save many human lives. There is even a glimmer of hope in that a breed of mice has been formed which can regenerate heart, toes, joints and tails and cells from such mice can be passed to other mice for help in their generation. (Sunday Times 28/8/05). How could the hope of amputees for regeneration be sustained other than by such experiments. `

During the 1950's antibiotics held sway in the Western world while attention in Russia was paid to the bacteriaphage and since the onset of bacteria resistant to many anti-biotics the West is also paying increasing attention to the former as a source of prevention. When a bacteriaphage can eliminate anthrax in hours then such a grouping is a must for consideration of future developments. As a class they are plentiful, very specific in their attacks and would seem to pose no danger to human beings. (Economist (8/5/04)

In one century we have survived to an increase in life from 50 to 75 and in our later years suffer diseases because our D.N.A. does make cumulative mistakes in its replenishment of our body tissues; as well there is the previously little known area where we have to cope with the present inability to cope with our necessary functioning. Even there we see signs of enormous progress in understanding their nature and this should further help healthy longevity. Time is always precious and the unnecessary delays caused by past thinking habits and resistance to change costs lives.

During the 1950's antibiotics held sway in the Western world while attention in Russia was paid to the bacteriaphage and since the onset of resistant bacteria the West is also paying increasing attention to the latter as a source of prevention. When a bacteriaphage can eliminate anthrax in hours then such a grouping is a must for consideration of future developments. As a class they are plentiful, very specific in their attacks and would seem to pose no danger to human beings. (Economist (8/5/04)

Energy

Domestic production and materials production take up some 30% of our energy usage. Transport takes up some 15%. It totals less than 5% of world G.N.P. Some 40% of energy is exercised in food production.

Ours is a very wasteful society. We hear of lean burn internal combustion engines and the use of renewable methanol and ethanol for improving the efficiency of conversion of energy. We are also told that were the internal combustion engine redesigned round these compounds, then in spite of the 20$ barrel of oil ,the cost differential of these renewable fuels would still remain a serious obstacle. Prices of oil in 2006 makes them promising! The ability to store energy relating to brake friction can also add its quota.

Here a main protagonist Brazil has bedevilled progress by producing misleading figures on such sources of ethanol and indicating its purpose to avoid use of foreign currency whatever the efficiency. There is also the promising prospect of converting waste products such as straw, corn stalks, and much other debris arising from agriculture, but such is some years away. (Economist.com 19/2/06)

Sugar Beet and potatoes too are potential sources of ethanol. The interest of such developments is that while such conversions produce greenhouse gases, the carbon dioxide produced could be re-absorbed by reaction with the chlorophyll in the next crop. There is the report that with the use of enzymes to degrade cellulose from agricultural waste to ethanol, there opens up the possibility of ethanol at 5 cents a gallon in the not too distant future, but agricultural space would be important if such a match to oil was confirmed. (Economist 1/5/04) and Stott tells us in the "Times" of 12/4/05 that to supply 1/65th of Britain's energy needs i.e. 1,000 megawatts, would require an area of seed rape covering the entire Scottish Highlands.

"The Economist "of 14/5/05 does confirm further progress in producing ethanol and biodiesel products, spurred on by a range of $40-$55 per barrel. The present offerings are 5%-10% additions although four million cars in the U.S. can run on 70%-85% ethanol and in Brazil some 30% of cars run on fuel containing such a level. The raw material is maize, sugar, rape oil, peanut oil, switch grass and in Canada there has commenced production based on straw. At the present price of oil some of these sources could offer serious competition but not at the foreseeable future if the oil price dropped to say $20 a barrel. However in meaningful usage the price of these raw materials could rise with demand, and there is as already indicated a problem of acreage. Before we panic as to available oil one should recognise that vast areas in Siberia and the Middle East have not been investigated for their potential.

The many alternatives are present in vast quantities but are too expensive to compete with oil at 20$ a barrel and oil has been less than this figure. At 40$ a barrel, tar sands clicks in with other sources and of course time should bring further efficiencies in production. Even corn based ethanol comes in at a base 60$ a barrel (Economist 22/4/06) and serious attention is now being directed to ethanol from agricultural waste.

However interest is still alive in renewable resources of energy as witness the focus in 2001, of the British government's interest in elephant grass. According to the "Sunday Times" of 28/8/05 that 800,000 acres are to be converted for generating energy in power stations, but subsidies will still be necessary for the farmer and the specially designed power stations.

The pursuit of efficiency still continues in many directions. Siemens recently advertised an improvement in generating efficiency of some 6% where the previous conversion was at a level of 46% and this means an increase of energy conversion of some 12.5%. "The Economist" of 5/4/97 tells of super-conductors mechanisms capable of saving another 8% electric power. While we must obviously take into account payback considerations and the costs of present plant obsolescence, one can see that the road ahead offers the prospect of some jewels.

In lighting, the saving possible are now beyond 80%. The smaller electric motor used widely is already at the stage of 20% improvement and even greater possibilities apply in electric refrigerators and T.V. All these innovations have paybacks of less than 4 years. The incorporation of such would also lessen the output of noxious gases from electric generation. Now too there are the considerable advances in train with L.E.D. (light-emitting diodes) technology. They have a very long life and have no delicate parts, but are costly to produce. According to the Economist of 5/10/02 ten watts units, which are the most powerful available, can be grouped in say, 12 units to give 120W lighting and in near white light. At present they are twice as efficient as normal lamps but only 1/3rd as good as fluorescent tubes. To be competitive they need to be as efficient as fluorescent lamps so as to take account of the initial costs of these semi-conductor sandwiches.

In the U.S. buildings use 65% of the electricity, 36% of the total energy used and produce 30% of the greenhouse gases (Economist 4/12/04). Further attention in conservation of heat in buildings could save by American figures some 30%, equivalent to some 500 million tons of carbon dioxide per year again with paybacks of less than 4 years. While this amounts only to some 10% of the world figures if one thinks of other developments in train this figure of saving could at least be doubled. Design of building for best heat conservation may also have a contribution. Last let us add that through the latest developments in glass insulation, there are great savings and on a 4 year payback.

According to "The World in 1978" the world-wide use by man of energy was some 7.5 thousand million tonnes of coal equivalent, 90% from fossil fuels and 10% from flow, and atomic energy. Energy received from the sun is a hundred times this amount but of course that energy also contributes to maintaining our present levels of temperature and ambience in life etc. At $15 per barrel of oil this means £75/mtce. There are tidal waves, geothermal sources, hydropower and wind. In all these routes one must ever aware of the political element known as the "not in my backyard" syndrome.

In the last decade there has been a tenfold improvement in efficiency of these newer instruments and in the U.S.A. it has been suggested that in the major windy areas, these can amount to 10%. The modern generator is less expensive in that production costs 16% of that of a decade ago. They are faster to build (1 year) and more reliable than they were. (Econ. 10/3/01). But there are also not just questions of the scenic view, ecological considerations but attendant fears of possible electromagnetic and other field effects which may possibly attend their presence from which we or our domestic animals may need protective measures.

There were in 1998 some 42 wind farms in G.B. running 748 operational turbines able to meet the needs of 200,000 households. Against this fossil fuel usage could be much reduced.

As to cost "The Economist of 3/5/97 tells us that GDP for 1955, taken on PPP assessments is for the U.S. and Europe 15, $14 trillion and I suggest $44 trillion world wide. Were we to double our fuel costs by replacing fossil fuels by other sources costs would rise by 2%, just making an impact on a year's GDP growth.

However we are a society **that measures by percentage increases not arithmetical increments** and thus a tiny amount introduced in operations early on make a great contribution a little way up the line in manufacture and distribution of products and services.

In talking of energy requirements and the use of renewable energy, one must recognise that wind turbines kill birds, wave mechanisms disrupt marine life. But wind turbines would cover inner London for under 2% of Britain's needs.

Only solar power seems relatively harmless and non-invasive, but here the area needed would be a further 50%, unless enormous strides in efficiency were to arise.

Wind-power is now a potential large scale contributor to our needs, and there are geo-chemical sources, wave power and tidal power. These other alternatives could make worthwhile contributions were they adequately developed, but they could not yet aspire to roles of importance until the relevant production facilities are in being. According to the Economist 31/7/04 figures are given for wind power of 2.5p per Kw/hr, by the British Wind Energy Association, 3.7p by the Royal Academy of Engineering and an offshore cost 5.5p. The corresponding figures offered by the R.A.E. are marine 6.3, coal 2.3, nuclear 2.2 and gas 2.1. This takes into account that full wind use is only available some 30% of the time and also that the new designs in use are strong enough to work downwind. Better still the" Observer" 20/7/03, reports onshore wind costs at £20-£35 off-shore £24-£40 per megawatt. Then there are hydro sources considered available at 5.0p, landfill and waste at 3.5p and 3.0p respectively. We must first carefully evaluate the important range of costs involved as well as which protagonists are at which end of the scales offered. At to-days prices, we may on costs alone able to step into a new field offering a great choice as to source and away from fossil fuels and even nuclear energy. The problem of the "waywardness" of supply and storage is now being attacked. Again there is the development using turbines that rotate about a vertical axis which it is claimed enables an almost doubling of gain from the wind's available energy.

For conservation of energy already gained there is the fly wheel, the use of the potential energy of supply being driven uphill and the reversible fuel cell showing important improvements in efficiency. It must be understood that the British consumption in 2004 is some 330,000 gigawatts per year and this equates to nearly 200,000 wind turbines plus back up plant when the wind is not adequate or the equivalent figure of 30 nuclear power plants. The move to consider offshore siting at greater expense will be increasingly mooted to diminish rising social resistance to their use on land. If windpower generators have dangerous level of electro-magnetic field strength associated with them or ascribed to them then there may be resistance to their use even in water as affecting sea life. It must however be repeated that this still awaits confirmation of a source and its availability in quantity and at an acceptable price.

With respect to Britain and its target for 2010 this needs some 500 wind turbines per year each machine of 3mw capacity and they will need service roads, huge concrete foundations producing energy with present day machines at some 3 times that from current sources such as gas ,coal or oil. The generators will be a hundred feet high and 375 turbines could generate 850 mw adequate to service

800,000 houses, and still have less capacity than Sizewell. However the new machines 330 feet high with a potential of 59 mws each are entering the market.(Observer 20/7/03)

Problems will still rear their ugly head as in the destruction by storm of the water-wave plant in Bergen, and off Scotland reinforcing a land base for its essential equipment. A U.S. concept of buoys with a piston riding the waves is now being evaluated since it is less susceptible to storm damage.

For solar cells, while overwhelming energy enters the atmosphere for conversion, the process so far is very inefficient. Failing an improvement its use is likely to be limited to the most rural of areas for the midterm. New developments as revealed tell of proto-type solar cells which have a maintained level of 17% efficiency according to manufacturer, Sharp, in the Times of 1/10/05 and 15% would move it into the range competitive with fossil fuels but still at twice the price. It still costs 28p per Kwhr and is therefore still, more costly than fuel cells. However this figure is one-third the cost level of 10 years ago, and 1/50th the costing for 1970.(New Scientist 12/4.97). With the new prices of oil at $60 a barrel, a new justification could arise

However New Scientist (7/12/02) tells of a further development where the use of gallium Indium nitride as base almost doubles the efficiency of conversion to 50%, but unless the cost of production can be materially reduced it will never be competitive. A considerable niche area could nevertheless be developed where other sources need expensive infra-structure, or where we are willing to accept the increase in cost as a small enough sacrifice for the advantages offered.

The fuel cell (Econ. 1/7/200) is now receiving more attention since it is much more amenable to transport usage than other newer forms. Ballard of Canada has reduced the size of a cell suitable for a small automobile to the level of a micro-wave oven.

The cells do not need the close arrangement of the present system and can replace the modern combustion engine by a design 15 cms thick and as wide as the conventional chassis.

The problem is storage. There are cells which can use methane and methanol, (the latter, feasibly from renewable vegetation sources), hydrocarbons which can be stripped of their hydrogen, by a sodium boron hydride solution using ruthenium as catalyst. (The Economist 12/1/2002). Carbon based compounds can be stripped of their hydrogen with oxygen and steam. The process still leaves the full complement of carbon dioxide and one should try and use the intrinsic energy inherent in the carbon content during its conversion. Since it is produced in bulk it can easily be removed and at minor cost. Thoughts of storage in depleted gas and oil reservoirs also arise as well as taking advantage of the reaction of carbon dioxide with silicates to form carbonates.

Cells are down to one-third the cost ruling a few years ago and power has been boosted by some 60 per cent, the thermal efficiency at site now being more than 150per cent that of the petrol engine. Sited sources would seem to be able to offer generation at not much greater cost than today, and a range offering about 400 miles range is now attainable. The best storage of hydrogen is being tackled and forms of carbon offer promise. However hydrogen can easily form an explosive mixture. There are the problems that to the extent that diffusion enables escape, that hydrogen could seriously remove ozone from the upper atmosphere and that an increase in hydrogen availability would encourage the microbes that feed on it.

Siemens Westinghouse promise a price at 1000-1500d/kiwi by 2004 and there is the possibility that others will achieve 500d/kiwi. The major problem at this time is that the platinum required at the anode of the cell for the conversion of hydrogen and oxygen in this application would cause a scarcity in world supply.

Wind-power cannot always be used at the time of production and therefore is not always acceptable onto a grid. More recently Prof Fells is developing a fuel cell (I.O.S. 28/1/01) to store such and the combination may prove a breakthrough. There is a long way to go, and the conversion of these raw materials into energy still produces a like contribution to the Greenhouse effect, agreed without nitrogen oxides, it is just the site that has changed not the problem.

The Sunday Times of 21/1/96 tells us that the North sea oil area is seen as an increasingly prolific producer at a level equivalent to the Saudis. Indeed they confirm this view in their newspaper of 15/12/02, by telling us that there is more available in reserves than has been extracted since 1967 and estimates of what is still available are more than has ever been presented as estimates in earlier attempts. Nigeria is a place where oil causes greenhouse pollution on a scale greater than G.B. sending 34 million tons of carbon dioxide into the air by burning most of the fuel gas accompanying the oil. 4% is burned here. (Indep. 10/12/95).

Reserves of oil have "RISEN" some 20% over 20 years. Odell avers that since 1971 over 1,500 billion barrels have been added to reserves while in that whole period just over half of that sum has been consumed. The fact is that technology has allowed recovery to rise in that period from 20% to 35% which means that there still remains 65% for more advanced technology to exploit. There is no reason to doubt that further inroads will be made in time.

Today proven reserves have a forecast life of 66 years, but we have not felt the full bite of China and the Far East. Coal is there for some 200 years and technical advances are still in progress and 1/3rd of electricity comes from that source, despite a reduction in the U.K.in production from 280 million tons to 28. In 1972 The "Club of Rome" published "Limits to Growth". It indicated total global reserves of oil at 550 billion barrels, in 1979/90, yet 600 billion have been used and reserves of 900 billion found. There is error, improvements in techniques of exploration and recovery as well as licensing and tax deceits involved here, the agenda may have hidden facets. Odell offers as a conservative figure 5,000 billion barrels and no peaking until 2050, gas not peaking until 2090. (Sunday Times 24/12/04). World demand today is 81 million per day and rising.

There is added the interesting claim that if all the oil and gas reserves known to man were burned then the carbon dioxide level would only double that present before the start of the industrial revolution and if the coal were also eaten up it would still amount to only 4 times that level. (Economist 6/7/02) There are thus upper limits for the danger of carbon dioxide emissions.

Since we are not replenishing the world's oil supplies, and we use oil in vast quantities (80 million barrels per day) it cannot be gainsaid that oil as a source is diminishing. However, the rate is subject to many factors. The forecasts in the 1970's for 2000 have proved far too pessimistic. Today the average level of oil recovery is 30%-35% optimists see a return of 50%-60% and there is therefore the possibility of returns to previously spent fields. The most expensive fields need only 10$ per barrel to cover all costs including the finding and developing of wells. (Independent 3/11/01). Add to this modern techniques which have halved the cost of raising oil to $4 a barrel and finding and development costs reduced over 20 years from $20 dollars to $7 and prospects here do not seem desolate. (Independent 8/12/01)

For greening problems we have the developments of Omega Oil where a single shaft can be organised with underground lateral tributaries. More expensive initially, yes, but in an area of 8,500 acres only one pump is required against 200 etc etc, and thus maintenance costs are reduced, as is the staff and infrastructure and we could get twice the level of oil that is available today.

Backing these reserves are simply enormous reserves of tar sands and shales, extractable by known techniques at possibly $7/ barrel but needing investment will, and there is coal in its own right, convertible to oil at 3 barrels /ton.

The average car in its life produces 35 tons of carbon dioxide and there are (Econ. 22/6/96) some 500 million cars. Road transport contribute 1/5th of the carbon dioxide produced and uses half the world's oil. It is still necessary to power it using say methanol and the water vapour will contribute to the greenhouse effect. The recent interest in air traffic needs to focus on the fact that it produces some 3% of carbon dioxide rather than the 22% from road traffic and very little alternative exists if you wish to travel far afield in a short time.(The Economist 10/6/06)

A tidal barrage across the Severn proposed in 1924, could have an efficient life of 100 years , could supply 8% of electricity but offers energy at three times the present price. The uncertainties of the free market renders the life of a plant where capital return is reckoned in one or two decades or more, a non starter, yet who can promise that today's prices will be maintained in twenty or forty years. How many times have we looked at structures of the past and admitted that we could never "afford" the cost today.

Devices already available could save 25% of fossil fuels without any sacrifice, and in all probability much much more, and there is the likely offspin of lessened maintenance to our town structures etc.. There are even the possibilities already making good headway using heat from generators for the local populations and waste from man's cast-offs as a source of energy.

Water power can supply some 20% of our current needs and the vast assembly of capital to exploit the use of dams has taken the forefront of our thinking in this arena. For the investor and for the state recipient their very size has made them a question as much of prestige as use and has vitiated the arguments based on their real value.

It is said that the Aswan Dam is contributing to cloud formation in the area and raising the water table. The damage shows itself in the attack on their ancient monuments and also arises from the entrapments of salts and their build-up.

Elsewhere dams swallow forests, vaster quantities of carbon dioxide result, and whole tribes are forced to change their habitats and habits. The promises given of compensation are treated with the same unconcern as those involved in the displacement of the Londoner by the rail companies in the last century. There is also the possible problem that containment of supplies allows concentrations of an undesirable nature, of Weils disease and Cryptosporidia, as well as concentrating the political problems. The considerable increase in bilharzia arises from the inception of the dam and the consequent irrigation system.

Nuclear Energy and Radiation

Too often the pretensions of the "Greens" and other "do-gooders" have sidelined man's concerns, and acted in a manner inimical to progress and away from recognition of the need for an end to man's control over man.

Nuclear energy is an abandoned mess, thrown without consideration into a morass. It has cost much, raised many fears some of which are legitimate and if abandoned will need resuscitation at some stage, probably expensively. Amid the vociferous cries and hysteria raised especially against this source we needs must examine whether the "Greens" who see a wolf in every shadow have at long last found a wolf, one that could indeed overwhelm us. The question that is important is whether it can be domesticated or at least tamed.

Perhaps like Macbeth they hope to destroy this Banquo for their own ends, and it perhaps they have only scotched this dragon but not ended its life.

To some this avenue opens a veritable Aladdin's cave of resources while opponents stress the Pandora's box aspect and the devastating dangers that could arise from its development as a resource. Here indeed is an energy source with no contribution to acid rain, no changes in the extent and distribution of forms of life, and no noxious gases. But it does mean attempting to play safely with fire!

Among the important rules of the game which the protagonists on either sides wish to play to win, there is involved forces concerned with the cleanliness of the operation as indicated in the previous sentence. There are the possible consequences of Three-Mile island, a Chernobyl or even worse. There is also to consider the consequences of a war-like action destroying nuclear generators coupled with a situation where no corrective action becomes possible, either because key personnel have been eliminated or where access to relevant sites are made impossible. Chernobyl has highlighted to the uninitiated that the consequences could not easily be contained and that an effect could also be a fall-out on the perpetrators. Last but not least is the small but ever growing stack of spent fuel which can remain a danger for aeons of time. Even this danger can be reduced by conversion to elements of shorter lives, say hundreds of years, but at cost and a waste of the potential energy encased in the rejected product. (The Economist 18/3/06)

Of course improvements offering new and more efficient designs and modes of conversion are now available some really reducing the danger to very much smaller proportions, but one would still not wish to be in any danger of such a possible catastrophe. (The Economist 3/6/06) and in any case these improvements are for new reactors. However we must remember that other forms of energy are also not free of risk.

The reaction to produce energy by this means, like the ozone "hole" phenomenon, partakes of a chain reaction and such reactions if they pass a certain point cannot always be contained. There are those who tell us that the chances of danger is statistically small enough to be neglected, but statistics are what happens to others not to oneself. This statement is not made out of a wish to disparage statistics. Statistics are most important and relevant second line mathematics in the search after certainty, and often there is only statistical certainty. To be told that Chernobyl or worse cannot happen in these terms is no comfort, after all even with the claimed superiority of Western technology we still had Three-Mile Island and there seems to have been a number of near misses in Britain and at Harrisberg. A one in a million chance in the face of possible world catastrophe does not look so good when we learn that there are some 450 such in action, with more to come, and the forecast chance may have a range of uncertainty which could pale the cheek especially as we become aware how figures are "tailored" to satisfy the Powers that be. The Institute of Physics tells us that new designs can reduce the probability of core accident by a factor of 10 and produce 1/10th of the waste, even recycle that waste, but we have to live with those of previous designs for many years. Fast Breeder Reactors while more than twice as efficient in no way lessens the risk. Incidentally little importance seems to be attached to the dangers to those exposed in the mining operations, for the raw materials involved in this venture.

Insurance underwriters are always wary of a risk showing a number of minor incidents, always suggesting that statistically it brings the possibility of a major calamity to the fore. Where results, however unlikely, could involve near or total disaster on the scale or beyond that which has been made plain to us , pursuance of such a path is not warranted by humanity were it the only road and the only source of fuel. And there are alternative fuels! Nuclear energy in its present mode is not that cheap and its use is in many cases a state strategy against being held to ransom from within or without. It is for such a reason that Thatcher saw fit to break the coal mine industry. **It also gives access to atomic weapons.**

Teller in 1956 offered a general reactor design which was a safe reactor. A semi-scale model was designed some 3 years later. The aim was to secure that safety was inherent in the design, not by enormous arrays of instruments. It would not be reinforced by further banks of instruments to be able to correct malfunctions or non-functioning of previously activated arrays. The safety was not to be a function of the depth and the breadth of such arrays. It was to be safe as a result of the physics of the design.

The West Germans built an experimental high temperature cooled reactor designed with a similar view, as did Sweden in its design on a proto-type scale. They differed, but safety in all cases arose from the materials within the system. The designed reaction would absorb in good time the maximum run-away energy that could possibly be released by reaction and which could otherwise cause a Chernobyl and worse. The system itself was to be fool-proof even fail-safe. Any instruments for use could be added and obviously the result of a fail-safe system would contribute simplification and cost reduction in their use. The projected designs were smaller, the initial capital outlay somewhat more costly but still economically not out of range. It was not pursued. Economic considerations may have been present, but the main driving force at the time was to be in the forefront of earliest manufacture for a quick sale into a gaping market and development of existing systems was pushed ahead and these safe designs relegated to an off-shoot of particle physics research. There is however renewed growth of interest as nuclear waste from present designs is increasingly seen as a major problem.

Since then there has been an interest in developing fusion, indeed The Economist of 20/7/02 has mentioned $17 billion spent in the U.S.A., but the effort has not borne fruit although the fault may lie with the low priority given it. By an large it has been treated as a minor project in programs concerned mainly with sub-atomic particle physics, as a sacrifice of revenue from the main project and the promotion of individual reputations. Our present nuclear industry was built as a war project with no financial restraint a condition that no longer applies. But safety against "rogue " states etc may continue to pressure this work forward. After all it does promise manifold advantages.

As ever safety, paled into impotence in the face of a quick buck. The fact that while a typical plant operating today may cost $230 million and decommissioning is now assessed at $635 million still means that the cost is 8-9 cents kw/hr.

Today, engineering designs always incorporate a safety factor which is a safe experience multiplicand of forces, theoretically derived as likely to be encountered. This factor takes into account imperfections noted from experience as arising in processed material and the vagaries of stress during the projected life of the fabrication. With the developments of new processes and examination techniques, this will be altered. The factor is an important contributor to the cost aspect, especially capital cost. As a result 19 R.B.M.K reactors of Chernobyl design with improved safeguards will continue in the Russias. Pressurised water reactors as at Sizewell are safer, but are they safe? Gas cooled, helium cooled H.T.R.models are safer still, but they work under conditions where experience is not that great, and there is the addition to such of continuous flow of fuel in the form of pebbles already on trial for 6 years without mishap . But blockages in the feed mechanism could occur with such a presentation of material and a view as to helium as a coolant, escaping the pressures of politics has slowed the onward thrust of such efficiency improvements. Equipment by its very nature is a possible source of long term fatigue, corrosion, and vibration is eating out the heart of structures and instruments. There is even the personal element of decision.

The problem is to achieve sufficient input of energy to make itself self-generating by magnetic hold and keep it away from walls for more than the few milliseconds so far achieved. For this endeavour larger capacity machines i.e. more expensive machines , are necessary. " The Times" 10/11/04 does mention an E.U project to build a fusion reactor somewhere and further news indicates the country as France. "The Observer" 18/5/97 discusses a new development using a laser technique which by using up most of the inherent atomic energy in the fuel would be safer with respect to residue disposal and be fail-safe. The reaction would cease so soon as the laser initiator stopped. Its base material would be Thorium which is more available than Uranium. Unlike the original plants which had their derivation in processes to make atomic weapons this process would produce no such end products, and indeed could in major fashion transform existing waste fuel to forms much, much easier to treat.

As already pointed out elsewhere the costing of the different modes of producing energy should hardly be graced by such a term. In 2000 B.N.F.L. has raised their nuclear waste liability from £27 billion to £34 billion over 100 years. The discounted amount being £16 billion. The danger from waste can be limited to lifetimes but only when converted to short term life at considerable cost.

France is 70% nuclear. The French Government claims that the fuel obtained from this source is the cheapest they have, probably based on the fact that repeats rather than indecision as to type of reactor enabled economies to be more easily implemented. There is also the fact that the French state used its powers to override the objection lobbies and their delaying tactics, which cost the former C.E.G.B. millions in defensive measures and even more in the inevitable cost of these delays. Cynics have also suggested that French government subsidies never as yet fully disclosed has also helped. The Sizewell return of capital was based on a ten year discount rate and the French on a 20 year discount, and such purely accounting considerations do make all the difference in cost estimates, when supplied for economic approval. Sizewell has lasted 50 years. The vagaries of interest rates over the long periods before inception of plants and delays political and otherwise do play havoc with projected figures.

In the comparison of generation costs between different sources of energy we have the Economist of July 1988 telling us that it all depends---. They gave a theoretical example where with a discount rate of 5%, the net value of producing power is 45% above coal at the 1988 price of coal. At 10% discount, coal fired energy is the winner by a little. Coal plants, gas plants and nuclear plants take very different times to build. Where a nuclear application can take 3 years just to get past objections, a gas fired project can be completed well inside that time. The ruling average rate can certainly not be forecast within such figures. Hitherto coal has never been expected to pay for its waste products and there is little doubt that nuclear waste costs are continually fudged.

But we already have an array from which France draws 70% of its energy, there are the Low Countries all close to G.B. relying on similar levels. There are at some 450 around the world which in the above light may all be considered as unacceptable designs because of their potentiality, however remote, for making catastrophe. In England sheep farmers today tend stock they cannot sell, on land infected by Chernobyl, and France and the Low Countries are nearer than the Ukraine area.

At one time it was confirmed to all and sundry, that nuclear generation costs would be competitive. Then like today's Eurotunnel costs seemed to double and treble, and we do not talk of pennies. The fact is that energy is State directed and its direction varies with time and the politics of that moment. It is not motivated by purely economic considerations, whatever the homilies.

What should terrify us all, is not nuclear energy, but its use, where adequate controls may not be in place. This consideration obviously applies to Chernobyls as well as some 20 other deaths where employees subjected to radiation have died. We are also aware of current data on leukaemia in the young and possibly "present" in the sperm of parents working on the site. However "The Times" 10/6/05 indicates that after many years of anxious questioning there is no evidence of an increased risk of cancer close to nuclear power stations. Nevertheless, some risk was confirmed where reprocessing was done, as at Sellafield.

One needs to realise that gas sources are of relatively short span, that coal is becoming more and more expensive to mine even though still with enormous reserves. Besides, the coal mine is not the place anybody should wish to send their sons. Improvements therefore, in coal site gasification may also be worthy of pursuit. Oil is still available into the medium term, but the areas of concern are where they are large political and/or price tags. In this context there comes a requirement for alternatives which have little relation to so called free market principles. In this arena we have spoken of various possibilities which might in major fashion replace fossil fuels, perhaps almost totally ,but it does need a greater understanding than state politicking can serve up.

Sellafield has a prime role as keeper of existing raw materials for nuclear energy and reprocessing spent fuels to recover plutonium for energy and especially for nuclear weapons. Its existence allows the British State to produce these weapons without " a by your leave". These spent fuels need to have extracted from them uranium and plutonium before corrosion of their containers sets in, and the waste material (in world wide extent some 7,000 tons per year) contain active material with a life of medium and long term. Even material of lower radio-active intensity still needs a home. Taking into account the recovery process, what strikes the outsider is that in today's designs, more than 95% of the power potentially available is wasted. In fact because of this high present wastage, nuclear energy could become within a century or two a resource too scarce to use economically unless better conversion of the active components can take place and such necessary techniques are available. The new fast breeder reactors are now capable of extracting more than 60 times the energy previously obtainable and about half that inherently available. A snag is the cost and control of liquid metal coolants but the nuclear waste problem would recede.

Fusion, perhaps by the laser route remains the best hope of meeting possible objections, based on real considerations, but cold fusion is still out in the cold and other means are distant. Here again there are some legitimate safety fears in that while the particles that may escape may not be radio-active their great energy is capable of rendering the containing structures radio-active as well as weakening them. The problem then transfers to the storage of damaged or otherwise useless containers which may only have a life in use of a few years. With the massive concentration of energy involved ,any accident could still cause catastrophe even to the extent it were dealing with non-nuclear aspect. But we would be moving from the essential reaction as the danger to a lesser danger and it may well be possible to erect energy baffles which could successfully contain this problem.

Radio-activity is posited as the great danger, but only since the arrival of the admittedly dangerous twins, nuclear weapons and energy. Apart from calamity situations it is time to take a mature view and realise that ever since the world was formed, there was radio-activity present and the entry of man made little difference. Every rock, every plant, every breath is radio-active. Even since 1945, the era of nuclear energy and the nuclear bomb, any rational examination of the world we live in does not in any way alter this conclusion. While there may be some argument that a certain incremental increase in radio-activity may trigger off a much, much larger danger to life, we are met by other critics who put forward that the danger increases uniformly with the concentration and that as such there are no sensitising plateaus with respect to concentration.

In the top 6 inches of soil, one square metre in area there will on average be 2 kgs of uranium, 6 of thorium and one of potassium 40. The journey round an average domestic garden gives you 19 % of exposure. At sea level cosmic radiation is 25 millirea/yr at 9,000 ft it is become 90 units and in practice exposure of atomic plant workers is about 2 units per yr.

Examination of the Hiroshima event, although not quite comparable to continuous small scale doses indicate no heredity consequences, and of the 80,000 survivors only 300 died of radiation induced cancers. All other survivors are in good health allowing for age. The Sunday Times (17/3/96) indicates that the major problems found in Chernobyl were the result of the consequent break down of the infra-structure. There seems no increase in congenital deformity ascribable to the event which remained 3% of those born. Children suffered at 50-200 time the normal level of thyroid cancer and statistically 1/10th of these would die as a result. These levels have been attributed to the retention of radio-active iodine in the body. The feared caesium passed through the body quickly and was a lesser danger.

The article by Wilkie in" Independent" of 6/6/2000 finds similarly in respect of Chernobyl. The initial Reuters report indicated thousand of deaths and health damage to tens of thousands. Of more than 100,000 drafted in to clean up the site there has been 31 fatalities from blast, heart attack, thermal burns and acute radiation poisoning. And this has been the conclusion of 90 medical studies. Only in the case of thyroid cancers arising with children has there been a calamity and 299 have died. However even this figure could have been reduced had iodine tablets been used. 48 endangered species are thriving in that region because man as a predator is not active, of 270 bird species 180 are breeding and most of the remainder are migrants. Carp, pike roach and perch are thriving and there has not come to notice any birth defects in animals.

One is only left to marvel at the resilience of life forms to what are ostensibly disastrous conditions.

The Russians want as much aid as possible and the West wishes to minimise it for the same reason as well as to damp down fears on existing nuclear plants, and the cynicism that is justifiably widespread.

About **90% of the radiation** we receive comes from natural sources, from the sun and outer space. An air trip to central Europe carries a dose comparable to a years work in the nuclear industry. Wherever there are mines for radioactive materials and in the surrounding areas, there is radio-activity and we have high radio-activity in large areas of Cornwall, Devon arising from the soil, giving a dose some 100 times in one week, of that suffered by a worker in nuclear energy in one year. Tin mines, even coal mines are involved since they too will carry with their main mineral, concentrations of radio-active material. More recently we have been advised that radon at adverse levels have become evident in Somerset, Northants as well as Derbyshire. However the Independent of 8/6/02 tells of a research program covering children in 6,000 homes subject to radon in Devon and Cornwell where the intensity is some 3 times the national average and found that at these higher levels no adverse effects were attributable. We live what is termed a normal life span but according to the latest study reported in the Times of 22/12/04 radio-activity causes some 20,000 deaths in Europe and 1,000 in England every year. It causes 9% of lung cancer and 2% of all cancers in Europe. however it may at some stage be worthy of discussion as to the nature of the term normal and under what conditions a new level of normality could arise.

In the U.S.A. the picture is the same and there is little doubt that the areas concerned will be found to be even wider in these lands as well as the lands in the remainder of the world. The "Independent" has mentioned that radon seeping through from the earth is associated with the death by cancer of some 2,500 people in England per year. The total houses in these 5 counties deemed a risk has been put at 100,000 and these are based on standards put forward by the authorities. In America, the federal safety limit was set at 4 pico-curies of radon per litre of indoor air but when at this level the number of houses at risk were found to exceed 8 million, voices, important voices, whispered that at this level compensation could not be afforded. They saw the way to eradicate the problem by just legally redefining and raised the previously unsafe level to a safe level by increasing the safety threshold. All quite simple, no evidence needed, no work to be carried out except for that by the judiciary and state economists putting something on paper. The danger limit was therefore raised to 8 units and all became well with the pockets of the elect.

Two tons of uranium annually in a respirable form was until the 90's released from British electric generators using coal as fuel as well as less quantifiable amounts of mercury, arsenic and lead. But scrubbing could have greatly reduce these amounts.

In 1984 the estimated dose in the U.K. from all sources was 2.2msv, 1.9 arising from natural sources, 0.25 from medical exposures. If there is cause to be concerned with radiation why do we always point the finger at man-made causes, and would it be wide of the mark to assume that the approach is concerned with the anti-scientific ideology that pervades our teachings?

According to British Nuclear Fuels the earth supplies 14% of our radio activity, space 10%. Plants and animals can no more avoid these sources than we can and as a result our food contains some 12 % .We are then left with medical examinations accounting for 12% and some 50% arising from the contents of the walls of our homes. There is very little available from other sources.

There is greater danger in embracing a beloved, because we thus expose ourselves to 25% of the dose obtainable from the world at large. In the light of such fears even sex could make a lethal contribution. From all these sources of life today and from its beginnings, we have adapted to exist within these conditions. There is no sign of a critical level yet arising where man is unable to cope with the newer sources of energy. Indeed mutation by radio-activity may have contributed to our evolution and the primary process of gaining food and energy, and there is photosynthesis.

Thus the constant concern that man-made nuclear energy can contribute to leukaemia, general cancers, sterility has as yet little validity, except near re-processing plants where better protection is obviously needed.(The Times 10/6/05). One in 3000 dies by accident, there is little concern to give such a figure against the less established figures where natural background radiation deaths on slighter information, is one tenth that total. Man-sourced nuclear energy in normal use is far lower than this last figure. The leukaemia clusters gathered statistically relate to few samples and therefore the error bands become correspondingly wide. Whitehaven, which is north of Sellafield had a high level of leukaemia even before nuclear energy was brought to that site. It becomes even more arguable when as has been found, similar clusters are found in areas quite away from the nuclear dangers. At this level of enquiry one would do well to test whether there are any other likely local causes. French and American investigations do not verify any clear association and at this time one cannot be entirely cynical of their findings. Whatever the tremendous interests here involved, the ascription of disease to nuclear sitings may still be incorrect.

It is amazing how far forms of life can throw off the effects which surface as seemingly catastrophic. Greenpeace has pointed at Sellafield as causing hatching failures and deaths in the local population of bird and livestock including the near elimination of blackheaded gulls. It has now been indicated that the eggs would need 20,000 times the available dose for abnormalities to appear in the eggs and an even greater level to cause a significant reduction in numbers. There is the recent letter (Indep. 12/5/200) written by Danish leading scientific authorities countering the demands of a stop to Sellafield releases by their government minister. They stress that the level is insignificant in that the average yearly intake of fish from Danish waters gives a dose of 0.14 microsieverts/year and the average house gives 0.3 per hour.Chernobyl has shown little medium term effects on Siberian geese, and there was an end of the world scenario read by some on this event. In fact "The Times" of 4/1/05 tells us that despite pesticides, radiation , greenhouse warming etc that man has thrown at bird life twenty species of Britain's birds ,most of them migratory ,are establishing new longevity records . Perhaps they are even benefiting.

The Chernobyl escape still affects the sheep and farming of Cumbria but we ourselves have no figures to say whether a ban on their entering the food chain is justifiable or no. Certainly there was a considerable increase in radio-active concentrations at the time, world-wide. In all this we must stress that more work needs to be done because we accept that one death is one too many. The real do-gooders are seeking facts not exciting stories of horror. Evidently themes and totems matter more than people.

Now comes the news (Observer 6/1/02), that after 15 years it can be seen that the medical effects of Chernobyl radiation are less drastic than previously thought, though not to be lightly dismissed. Of 100 on-site emergency workers suffering radiation sickness. Confirming the previously mentioned findings in 2000A.D., 41 died, there was an increase of child thyroid cancer of 60 times in Belarus, 40 times in the Ukraine giving rise to some 18,000 cases. There was no leukaemia or other cancers and no deformities of babies by exposed pregnant women, at least not attributable to this disaster.

In this we also seem to forget that each wind turbine could also set up electro-magnetic fields to which domestic animals and man could be subject.

Radiation does not only constitute that arising from nuclear energy. Wherever there is a flow of electricity a magnetic field accompanies it. The higher the voltage the more manifest the field. This applies to pylons, but even enters the very dwelling houses and this phenomenon too has been adduced as a cause of child leukaemia. Of course electricity can be beamed through the air and collected from antennae ,but omitting pylons and cable would be costly. There is radio, television, the V.D.U. at work or play, sound waves and even using the telephone or pressing a light switch. All can expose one to electromagnetic energy. In this, one should not forget the micro-wave oven. It is now considered that a magnetic field can affect the rate of heart beat and even the ability of the mind to concentrate. Mobile phones also provoke a watchful eye but the level of interest and enquiry is not at a great level. Here the latest report from the

National Radiological Protection Board, (Economist 15/1/05)says "There is as yet no hard evidence of adverse health effects" and it supports the previous survey of 2000 A.D. They do however recommend units with low absorption rates but purely as a precaution. Their recommended level would be further reduced by increasing the number of masts but this would not be acceptable to the general population. Heat is generated in the brain by their use, but no data indicates whether such gives rise to problems although the present advice is to limit use for children.

There is even indication that fertility of eggs, even human foetal cells, can be affected. We also know that chemical reactions can take place as a result of sound waves and there is no a priori reason why we should not be affected by any sources of energy. It gets even worse when we consider how widespread is the use of hair dryers, electric blankets and what of these home ionisers? It may be that even the flow of water in pipes including central heating pipes plays an unrealised and adverse part. However we are still surviving and although investigation into these aspects might throw a further light, they are little likely to justify nail-biting tensions.

The latest stage in these investigations released in December 1999 is the work of Prof. Henshaw indicating a route by which the electro magnetic fields set up by high power lines could cause cancer. A generator in every garden powered by the wind is therefore hardly acceptable unless shielding can be designed. There is also the prolonged large scale survey released in 2001 which suggests that prolonged exposure at high levels could double the risk of a child developing leukaemia. Only 300,000 persons are so exposed at that level which is above 0.4microtestas and the average home suffers a level of 0.1- 0.2 units. The increase in expectation is from 500 cases of child leukaemia per year to 502. Whether this is a cause is not yet proven and it is known that this phenomenon does not directly alter the D.N.A. structure which leaves the cause of this marginal increase still in some doubt especially since 80% of the children exposed at high levels are not near overhead power lines. This should not end the matter but one must bear in mind that, as with the silicone breast implant claims, the endeavours are more often concerned with different agenda on both sides and truth is of little interest.

The fears aroused by the above while not as yet arousing the same concerns as nuclear energy, may all make contributions. V.D.U's have been mentioned with respect to pregnancy dangers and as giving rise to skin complaints and it may be that we should take note and attend to the removal or diminution of the ill-effects. Even office air-conditioning has been implicated as contributing to illness. Were these factors shown to make a contribution and they might well, one can choose to reject these aids, but it is necessary to note here, that the stresses of life in society ,today, may be a major contributor and this needs airing above whisper level.

Alternatively one needs accept that man has found nature at best neutral and has spent time amending, as far as possible, the external world to his needs while protecting himself as far as possible from the worst contingencies as foreseen. There are many who would suffer the untold, rather than deprive themselves of the facility of a telephone. There are few advances that have no adverse features and it is the overall balance which should contribute to decision.

Were the quality and validity of the work done applied to any less widespread and important a sphere than the air we breathe and the radiation we receive, it would not have deserved that much attention. More likely it would be put on hold and further information required before any of the claims made would be granted as substantive. However here we are not dealing with a drug or process of manufacture which at a disadvantage we could live without and which however painful or widespread could be contained. The problem arises because we cannot isolate ourselves as a species from the earth, the heavens or the waters. Here we have evidence of a steady increase in items which may be deleterious in minor or major degree to the life of mankind. The information suggests that man is making a contribution to the deterioration of his habitat, but it may still be a puny contribution. But it is all we can control. It may or may not tip the balance, we do not know. We need to know, just in case we can by our own efforts design plans for remedy. Meanwhile the changes envisaged may well become recognised as in largest part a positive contribution to our well being. It is up to the ingenuity of man to develop features which would allow mankind to enjoy the benefits without fear of the exactions of nature. **We need to know!**

The problems here arise from world sources. They cross all boundaries, and states are far from benevolent societies. The solution must therefore be of a global nature!

Endeavour is necessary to seek alternative means for ameliorating or redressing the serious problems raised but there is no way back even if desired, nor is a stand-still available. The way forward may always spring unpleasant surprises, but we have survived this far.

The important problems are increasingly global in natureand can no longer be rectified on the basis of individual State advantage. We have discussed them above. There are the means in a freed and growing technology to allow adequate solutions to the problems already foreseen. A socialist world society would free the best means to activate the solutions in this field. Awaywith being trammelled by the profit motive. We should see it as a compulsory zimmer frame to encase an otherwise healthy unit.

Conclusion

This trawl and analysis of the available data illustrates quite clearly that we are not anywhere near the catastrophe or the doomed future projected by the Green and associated movements. Their scientific pose is but a fig leaf hiding their conservative status quo outlook. They repeat but in a more scientific language, and limited scientific content the rationale, that man is forever doomed on the morrow and by his own hand alone, which message the human race has ever held close to its breast.

WE HAVE DEALT WITH THE EPHEMERAL NATURE OF MUCH OF THEIR CONTRIBUTION.

The power of man may now be great enough to make impact, but we are veritable pygmies in comparison with the deeds of nature. A volcanic eruption, earthquakes, changes in the suns energy including its temperature and the orbit of the earth over time makes a greater contribution. But it could be the case of the straw breaking the camels back or no. Let us therefore investigate but in the spirit of scientific facts rather than the fears raised by a code which does not seek to assess, and is only concerned to arouse irrational fears.

The human race is still in control and can regulate itself. A wish for a relevant increase in carbon dioxide uptake by plants would cause no problem that was insoluble, and so can all the other problems now thrown up as offering irretrievable disaster. Indeed were it made possible, rather than attempting the status quo, control of the Greenhouse effect is of greater merit for improving our lot. It could be beneficial to later generations by reducing the effects of a future ice age. The only area which may need some extra care is in the Ozone problem.

Unfortunately the State, with the "Greens", also has a desire for the status quo and with its members largely unable to assess valid science they make excellent bedfellows. The real problem is the lack of education in science and the consequent uneasiness of the population as to its nature and possibilities. Today in addition there has been an explosion in the level of possibilities at the personal, as well as the ecological level presented by a media little concerned with the realities of the situation, concerned only to startle, dismay and exaggerate.

The treatment here presented is meant to help dispel the mists surrounding the potentialities offered by our advances and save us from the Luddite. The West still needs the advances in train and beyond, and for the Third World it is a dire necessity and far beyond the dribbling charities which their betters offer so proudly. Even these amounts are largely consumed before any appears near the lips of those that need it most. Over the past 40 years a trillion dollars have left the hands of givers. but African governments, local business men and agents receive the most and there are many NGOs in the game needing expensive administrative support.

Bibliography

Environmental Economics	A. Cottrell	Edward Arnold Ltd.	1978
The Resourceful Earth	Simon et al	Blackwell	1984
The World in Crisis ?	Various Authors	Blackwell	1987
The Tropical Rainforest Ecology	D. J. Maberly	Blackie	1992
Thinking about the Future	Various Authors	Chatto & Windus	1973
Resources and Man	Various Authors	Nat. Acad. of Science	1969
A Chilling Scientific Forecast of a New Ice Age	F. Hoyle	New English Library	1982
Nature and Industrialisation	Various Authors	Oxford	1979
The Environment and Society	Various Authors	Hodder & Stoughton	1991
Green Britain or Industrial Wasteland	Various Authors	Polity	1986
Developed to Death	T. Trainer	Green Print	1989
Industrialisation and Culture (1830-1914)	Various Authors	MacMillan	1970
Climates of Hunger	Bryson & Murray	Univ. Wisconsin	1977
Agriculture : the Triumph and the Shame	R. Body	Temple Smith	1982
British Population Growth (1700- 1850)	M. W. Flinn	MacMillan	1972

Title	Author	Publisher	Year
The Economic History of World Population	C. M. Cipolla	Pelican	1972
The Political Economy of soil erosion in Developing Countries	P. Blackie	Longman	1989
The Threatening Desert	A. Grainger	Earthscan	1990
Development in the Third World	M. Morrish	Oxford	1987
Global Environment Issues	Various Authors	Hodder & Stoughton	1991
The Farmers of Old England	E. Kerridge	Allen & Unwin	1973
Alternative Technology	D. Dickson	Fontana	1977
World Hunger: 10 Myths	Lappe & Collins	Inst. for Food & Development	1980
Environmental Economics	A. Cottrell	Arnold	1978
The Inaccessible Earth	Brown & Mussett	Allen & Unwin	1981
The Evolution of the Biosphere	M. M. Kamshilov	Mir	1976
Mankind at the Turning Point	Misarovic & Pestil	Hutchinson	1975
The Roots of Modern Environmentalism	D. Pepper	Routledge	1989

Man Remakes The Earth

www.ingramcontent.com/pod-product-compliance
Lightning Source LLC
Chambersburg PA
CBHW020438220526
45464CB00002B/764